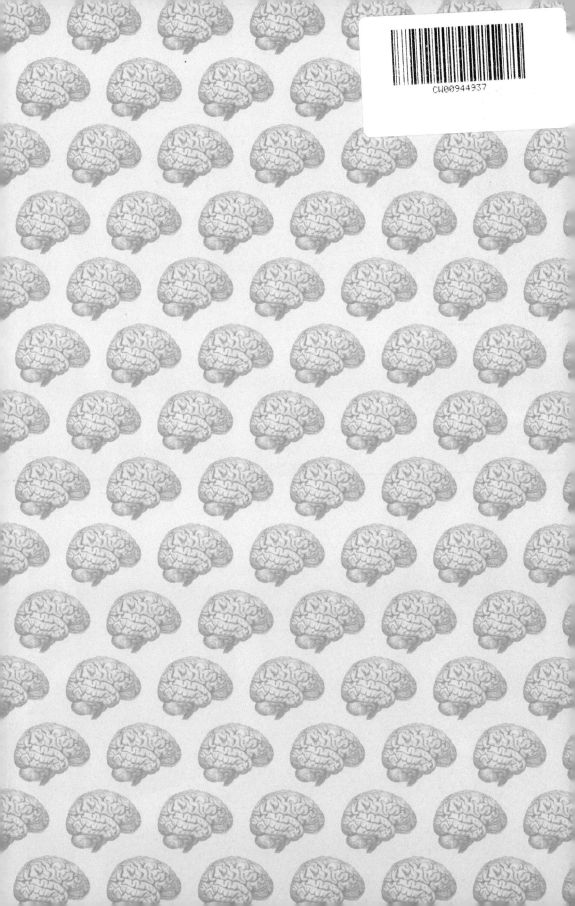

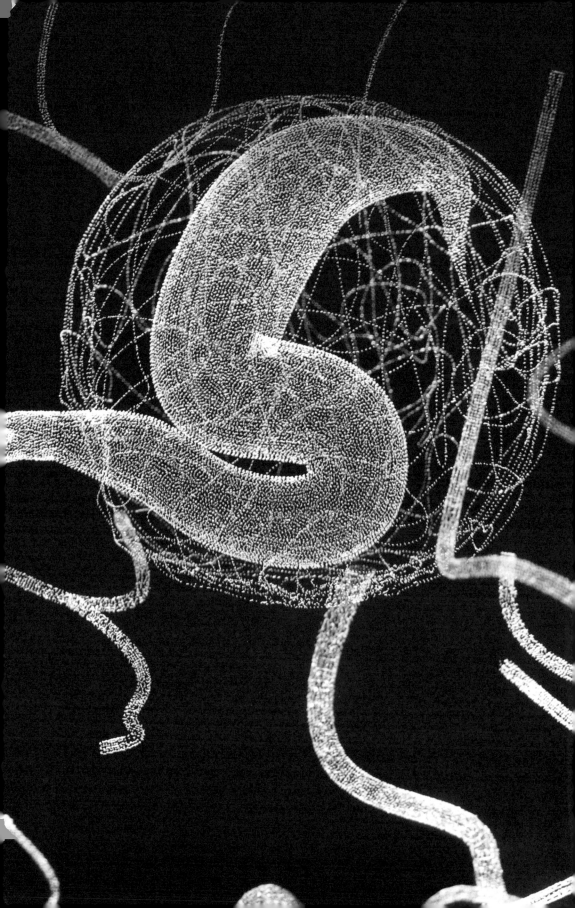

BRAINS

The Mind as Matter

Marius Kwint and Richard Wingate

First published in 2012 by Wellcome Collection, part of
the Wellcome Trust, 215 Euston Road, London NW1 2BE.

wellcome collection

www.wellcomecollection.org

Wellcome Collection is a free destination for the incurably curious devoted to exploring the links between medicine, life and art. Several galleries, a range of events and the unrivalled Wellcome Library root science in the broad context of health and wellbeing.

Wellcome Collection is part of the Wellcome Trust, a global charitable foundation dedicated to achieving extraordinary improvements in human and animal health. The Wellcome Trust is a charity registered in England and Wales, no. 210183.

Copyright © Wellcome Collection 2012

10 9 8 7 6 5 4 3 2 1

The moral right of the authors has been asserted.

All rights reserved. Without limiting the rights under copyright reserved above, no part of this publication may be reproduced, stored or introduced into a retrieval system, or transmitted, in any form or by any means (electronic, mechanical, photocopying, recording or otherwise), without the prior written permission of both the copyright owner and the publisher.

A CIP catalogue record for this book is available from the British Library.

ISBN 978-1-78125-040-2

Commissioning Editor: Kirty Topiwala
Managing Designer: Marianne Dear
Assistant Editor: Tom Freeman
Wellcome Library and Wellcome Collection photography: David Sayer and Ben Gilbert
Research Assistant: Sara Perry
Layout Designer: Allan Somerville

Printed and bound in Great Britain by Quadracolor, Surrey

Any omissions and errors of attribution are unintentional and will, if notified in writing to the editor, care of the Wellcome Trust, be corrected in future printings.

Headache (detail) (cover)
Helen Pynor, 2008
C-type duratrans face-mounted on glass

'Headache' is from the 'red sea blue water' series of photographic works depicting traditional remedies associated with particular organs. Pynor, who trained as a biologist in Australia before becoming an artist, is mindful that life originated in water and that bodies are mainly composed of it. Here it offers a metaphor for the combination of biological, historical and cultural processes that support the human organism.
COURTESY OF THE ARTIST AND GV ART, LONDON

Memory of a Brain Malformation (detail, see page 114 for full work) (previous pages)
Katharine Dowson, 2006
Laser-etched lead crystal glass
COURTESY OF THE ARTIST AND GV ART, LONDON

Contents

6 **Preface**
Ken Arnold

Essays
8 Exhibiting the brain
Marius Kwint

22 Examining the brain
Richard Wingate

32 Photographic essay: Brain collecting
Daniel Alexander

40 **Chapter 1**
Measuring/Classifying

56 **Chapter 2**
Mapping/Modelling

92 **Chapter 3**
Cutting/Treating

122 **Chapter 4**
Giving/Taking

160 Further reading and acknowledgements

Preface
Ken Arnold
Head of Public Programmes, Wellcome Collection

Each of us walks around with about 1.4 kilograms of soft, pudding-like tissue perched at the top of our spinal column. Closer inspection reveals an extraordinary number of constituent cells with a set of interconnections that makes each brain unique. We have gradually come to define life in terms of our brains being switched 'on'; turning them 'off' quite simply kills us.

From this bewilderingly mysterious organ emerges the totality of our experiences, feelings, ideas and understandings. Memories are made and stored there, habits formed and broken, personalities shaped and shattered. Little wonder, then, that the investigation of this most precious thing has been so central in attempts to fathom the essence of the human condition. But as Richard Wingate reminds us, the brain's "inner workings are completely inscrutable", with "no obvious moving parts, no pumps, valves or tendons". To bypass this seemingly intractable problem, much brain research has for centuries adopted a core philosophical assumption that the brain is a 'thinking machine'. Calculating engines from the 17th century and the first programmable computers invented in the 1830s served as early examples of that analogy, with the internet providing the latest model to latch onto.

The essential medical conundrum of studying life via death – not being able to witness our internal workings without stopping them – led to a range of audacious investigative strategies. One such tried to relate brain sizes and shapes to the supposed characteristics and interests of their owners. To this end the brains of the famous, and especially of the famously brainy, were studied very closely. Another method focused instead on partially damaged specimens, which gradually enabled researchers to elaborate the theory that mental capacities such as memory or touch resided in particular areas of the brain. Through the 19th century, the anatomical idea of looking at bits of the brain was pursued in ever-finer detail. Eventually, tissue examined under the microscope yielded the discovery of 'thought cells': the neurons.

This book (like the associated exhibition at Wellcome Collection) is stubbornly concerned with the brain as a venerated physical object. It investigates how brains have been measured and classified, mapped and modelled, cut and treated, given and taken. And despite the myriad means by which functioning brains can now be examined non-invasively, it asserts the continued relevance of pondering not only what the brain does, but also what it is.

Head divided into four cerebral lobes: profile
Bernard Hollander, 1895–1905
Pen and ink drawing with watercolour on board
WELLCOME LIBRARY, LONDON

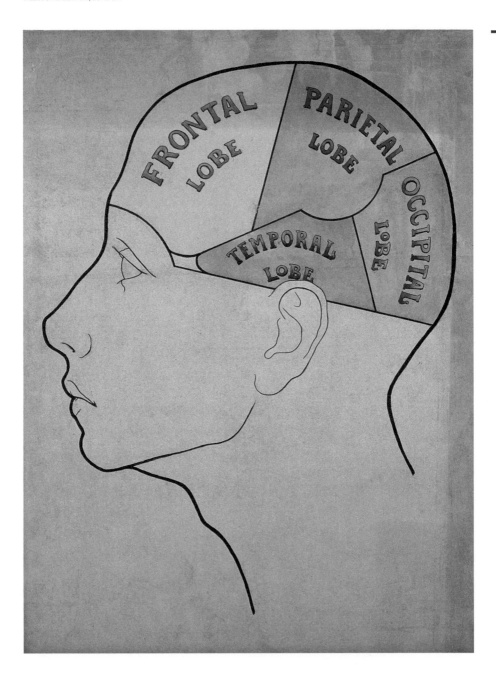

Exhibiting the brain
Marius Kwint

Neuroscientists estimate the brain to have an unimaginable 100 billion nerve cells with 100 trillion synapses or neural connections. There are numerous books, exhibitions, websites and television programmes about how these nerves and synapses combine to provide consciousness, thought, memory, movement, sensation, perception and emotion – what together is known, with connotations of authority and control, as the 'mind'. However, rather than attempting to explain what the brain does for us, this book explores what we do to brains. People do, and have done, a lot of different things to brains: we drill holes in their protective layers; we penetrate them with electrical wires and pass currents through them; we drug them; we subject them to intense magnetic fields; we cut out quite large parts of them; we shoot them out with bullets and scoop up the spillage with rubber-gloved hands and post the photos of this activity on the web; we slice, pickle, freeze and entomb brains in wax and plastic and then arrange them neatly in drawers and on shelves; and we make pictures – lots and lots of pictures – and models, as well as writing and talking about them. And that's only counting human brains.

It's customary for those writing about the brain to remark on the unpromising nature of its tissue in contrast to the richness, even infinity, of what it provides: 'the soul in a bowl of curds' and so on. Upon encountering a whole, real human brain, the sense of paradox at the cold flesh in hand versus the knowledge of consciousness and personhood can indeed be intense and overwhelming. "I cradled it like a baby," recalls the artist Susan Aldworth after observing a dissection at a London brain bank, "protective, respectful of holding 'someone' – someone very vulnerable in this disembodied state." Many studies pay lip service to the striking materiality of this object before proceeding to set out the brain as a subject, furnished with telling diagrams and seductive scanning images that force the substance of the brain into an imagining of its interior workings.

What if we refuse to go down that route, and instead allow the brain to retain the status of an object, with which humans engage in varied but purposeful activities, almost as if it were wood? To view the brain as amenable to different crafts and technologies reveals something of the actual operation of the brain upon the world. The aim of this book, and of the exhibition on which it is based, is to produce a physical encounter with the brain in its

Headache (detail)
Helen Pynor, 2008
C-type duratrans face-mounted on glass
COURTESY OF THE ARTIST AND GV ART, LONDON

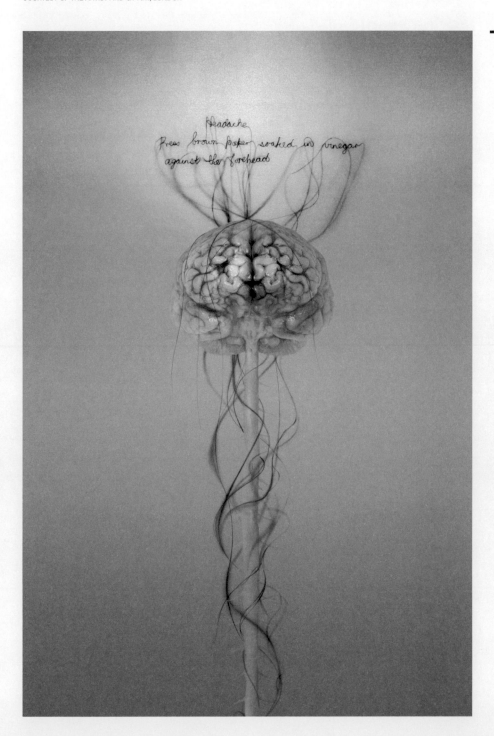

many forms, and to exhibit some of the apparatuses and methods with which it has been worked. The perspective stays for the most part firmly on the outside of its wrinkly surface.

Exploring the brain
Having said that, curiosity about how the brain works has motivated most of the ways in which it has been exhibited and visualised over the centuries. It is worth considering a few historical landmarks in Western concepts of the brain (although European cultures by no means had the monopoly on theories about it). Aristotle famously underestimated the brain and identified the heart as the primary seat of consciousness and the emotions, partly because the brain's delicately gelatinous tissue was so hard to dissect and quick to decay behind its armour of cranium and meninges. By the time of the Roman Empire, there was an established consensus that the mind (aka the soul) was primarily located in the brain, but in the second century CE, Galen – doctor to the gladiators, with plenty of broken heads at his disposal – placed the mental faculties such as reason and memory in the brain's fluid-filled cavities, or ventricles, rather than its solid matter.

Throughout the Middle Ages, human dissection was heavily restricted in Europe and the Mediterranean world, and most surviving drawings of the brain from this period are diagrammatic.

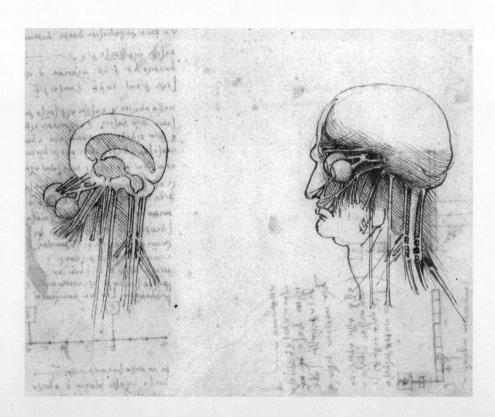

Diagrams of the head showing the eyes and brain with internal ventricles (left)
Leonardo da Vinci, c.1508–10
Pen and ink over traces of black chalk
THE ROYAL COLLECTION, © 2012 HER MAJESTY QUEEN ELIZABETH II

Coordination of muscle and visual mechanisms by means of pineal body (below)
René Descartes, *L'Homme*, Paris, 1664
Wood engraving in book
WELLCOME LIBRARY, LONDON

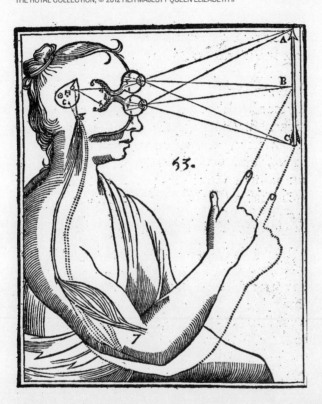

On the question of the ventricles, Galenic doctrine was still broadly accepted by Leonardo da Vinci during his furtive explorations of mouldering corpses some 14 centuries later. By the mid-17th century, the French polymath René Descartes argued instead that the small pineal gland, deep within the brain and named after a pine nut, was the regulator of zephyr-like animal spirits, analogous to hydraulics, which connect with the nerves and muscles throughout the body.

Scholars were able to debate the brain more confidently from the Renaissance onwards, when European university cities instituted anatomy theatres for the public dissection of cadavers. Working in 16th-century Padua, the great Flemish anatomist Andreas Vesalius published his unparalleled atlas of the human body. His engravings scaled new heights in the naturalistic depiction of the brain as an organ, with its graphically convoluted cortex and hemispherical structure. He dismissed the ventricle theory on the grounds that other animals also possessed that feature but only humans had the gift of a divine soul. However, conventional dissection techniques

Human brain with cerebellum and nerves (below)
Thomas Willis, *Cerebri Anatome*, London, 1664
Copperplate engraving in book
WELLCOME LIBRARY, LONDON

George Combe lecturing on phrenology to a large mixed audience in his Edinburgh home (right)
L Bump, after J Lump, London, 1826
Coloured lithograph
WELLCOME LIBRARY, LONDON

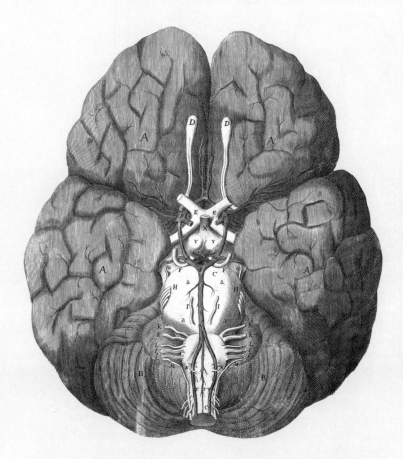

only enabled him to show the brain seated in the cup of the head, whereas in Civil-War Oxford a hundred years later, Thomas Willis, a critic of Descartes's pineal-centred account, managed to prise the brain intact from its vault and to discover the failsafe system of arteries on the underside that bears his name: the circle of Willis. His 1664 book, *Cerebri Anatome*, established neuroscience as a specialised discipline and boasted engravings by his friend Sir Christopher Wren that showed the brain in splendid isolation.

To the religious sceptics of the Enlightenment, the study of the brain promised a purely materialist explanation of human nature and variation, and so it became a guiding figure in the advance of scientific rationalism in the succeeding centuries. This single organ was scientifically venerated as the crowning glory of the evolutionary process that had, according to elite opinion, culminated in the global triumph of Western civilisation. Sheer intelligence became a bourgeois cult, for which the image of the liberated brain became an icon, its cortical folds bursting forth from their cranial prison with untold intellectual potential. Here was the Lockean man in charge of his own destiny, with a soul which, according to the latest scientific findings, comprised nothing more than organically generated electricity. The experiments of Luigi Galvani and Erasmus Darwin had, after all, produced the appearance and movements of life in the electrified corpses of frogs and recently executed criminals – a concept that notably fascinated and horrified the young Mary Shelley, eventual author of *Frankenstein* in 1818.

Enlightenment brought about many other bizarre ideas, among the more notorious of which was phrenology, the commercialised science that purported to read the character of a subject from bumps around the head. Phrenology appropriated the resurgent (and to this day predominant) theory of localisation, which holds that particular mental functions and faculties operate in specific areas of the brain, sometimes alone, sometimes in concert. Much more respectfully viewed by many 19th-century anatomists was the work of the Italian criminologist Cesare Lombroso, who revived the classical discipline of physiognomy by arguing that moral propensities were evident not only in the shape and expressions of the face but also in the dimensions of the brain that lay behind it.

Cabinet of 'Brains of Educated Orderly Persons'
Burt Green Wilder Collection, 1890s
Black-and-white photograph
DIVISION OF RARE AND MANUSCRIPT COLLECTIONS,
CORNELL UNIVERSITY LIBRARY

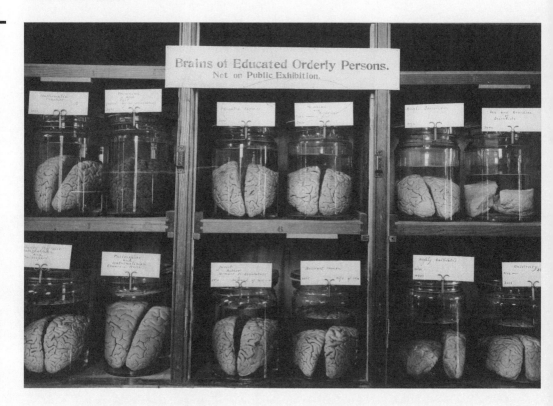

Collecting brains

Equipped with chemical preservation methods that allowed the indefinite storage of entire brains, a few ambitious anatomists such as Burt Green Wilder at Cornell University in upstate New York set about systematically collecting them for the purposes of anthropometric comparison. Brain-collecting was only one facet of the great classifying endeavour to which Western knowledge-gatherers were then subjecting the entire globe. At the Wistar Institute in Philadelphia, Edward Spitzka began drawing up a list of the brain weights of eminent persons, in which both the top and bottom positions would be occupied by famous writers: Ivan Turgenev at 2012 grams and Anatole France at 1016 grams. Biographical accounts appeared in learned journals, correlating the qualities of well-known personages in life with the properties of their brains in death. By the twilight of the 19th century, staining techniques that revealed the beautifully dendritic, or tree-shaped, structure of brain cells began to indicate how brains functioned at the microscopic level as a network of interconnecting neurons. The subdiscipline of 'cytoarchitectonics', or the division of different areas of the brain according to its different regional cellular types, was then developed by Korbinian Brodmann through a series of painstakingly constructed maps of the entire cerebral cortex in the 1900s.

With newly rigorous methods to hand, anthropometric collections continued to thrive in countries ranging from the USA to France, Japan, Russia and Sweden. The brains of many deceased greats, including Lenin and Einstein, entered the histological pantheon in the hope of identifying the substrate of genius, and indeed it became quite a fashion for men and a few women of science to bequeath their illustrious grey matter to their colleagues – a story engagingly told by Brian Burrell in his book *Postcards from the Brain Museum: The improbable search for meaning in the matter of famous minds* (2004). However, as Burrell's title suggests, few enduring conclusions were drawn, except to cement in the public mind the prejudicial fallacy that human intelligence was fixed, innate and directly correlated to brain size. This outcome was somewhat against the intentions of progressives such as Wilder.

"Sheer intelligence became a bourgeois cult, for which the image of the liberated brain became an icon, its cortical folds bursting forth from their cranial prison with untold intellectual potential."

While the powerful have usually been able to exercise their will over their post-mortem fates, there were many more instances in which certain types of brain were held to be of greater scientific or symbolic value than the wishes, or even the lives, of those whose heads contained them. Under traditional arrangements for the anatomisation of executed criminals and the unclaimed dead, many had had their cerebral matter, along with the rest of their bodies, denied the sanctity of the grave. And under the Third Reich, thousands of alleged mental defectives were even killed expressly for their brains under so-called euthanasia policy, culminating in the Aktion T4 extermination programme. Many of the physicians who had supervised these mass murders in the name of genetics went on to perfectly respectable postwar careers in both East and West, their papers still being published, and it was up to recent generations to bring light and some semblance of restitution to this episode.

Whereas the Nazis had the temerity to pursue their colonialism on the European continent, the Great Powers had been doing so for decades throughout the so-called darker parts of the world, and their scientists had been collecting, classifying and measuring apparent brain capacity all along. In most colonial settings, however, fresh brain material could not for the most part be readily obtained or preserved, so European researchers made

Frankenstein film poster
1931
Coloured lithograph
COURTESY OF NBCUNIVERSAL ARCHIVES AND COLLECTIONS

do with skulls. They would frequently dig up the burial sites of subject peoples to measure skull volumes and proportions. They also diligently photographed and applied headspanners and gnomiometers to the heads of their living, and often resentful, subjects.

Eugenics and anthropometric collecting fell into disrepute after World War II, but the latter bequeathed to popular culture the science-fiction image of the brain in a jar, as if suspended in an artificial amnion awaiting some ghastly rebirth. The most influential of these was the 1931 film of *Frankenstein* with Boris Karloff, where, in a shameless travesty of Shelley's novel, a blundering assistant supplies the creature with the brain of a criminal rather than the perfectly formed specimen that Dr Frankenstein intended. As a result, Brian Burrell says,

> ... there are now two monsters – Mary Shelley's and Boris Karloff's. One is an athletic superman who talks incessantly, while the other plods along and never speaks. One is driven to murder by his fragile ego, the other by a defective brain. One is wounded by careless nurture, the other by cruel nature. Shelley's monster imposes his will on a world that rejected him. His modern counterpart, on the other hand, has no will at all. He is simply a prisoner of his misshapen brain.

Burrell explains how the film's scriptwriter, Robert Florey, was pressed by the producer, the 21-year-old Carl Laemmle Jr, to come up with a clearer reason for the creature going berserk than what was offered by Shelley's humane critique of scientific hubris and popular prejudice. Florey inserted the idea of the abnormal brain (or "*dysfunctio cerebri*") and "[t]he rest, as they say, is cinematic history". Television offered further fantasies of the disembodied but animated brain, for example the wired-up cortex of an intergalactic villain in the 1976 *Doctor Who* series *The Brain of Morbius,* and the character of Big Brain – just a big brain – in Matt Groening's cartoon *Futurama*. Similar ideas informed the 'brain-in-a-vat' hypothesis discussed by the Harvard philosopher Hilary Putnam in 1981. In this famous thought-experiment, an update of Plato's Allegory of the Cave, Putnam reasoned that he could not be a mere brain connected to a supercomputer providing the illusory sensation of volition and perception. All these scenarios played on the biologically absurd but ideologically plausible notion that humans are ultimately reducible to their brains.

Today, anatomical collections mostly gather dust amid the smell of formaldehyde in the basements and backrooms of medical schools, unless they are presented to the public as historical artefacts. But modern brain banks are now burgeoning around the world, furnished with super-cold freezers, opaque plastic buckets rather than jars, vast computer databases as well as walls of slide-drawers, and new techniques for the genetic analysis of tissue. These are the means by which researchers endeavour to shed new light on the epidemics of Alzheimer's, Parkinson's, multiple sclerosis and other neurodegenerative conditions that especially threaten the future of affluent societies and their ageing populations. Ordinary citizens, whether healthy or afflicted, are being asked to donate their brains to science, under strict regulations to ensure that dignity and consent are respected. In the field of brain-mapping, delicate and sophisticated technologies and production techniques are now engaged in compiling huge online atlases of neural structures and connections, sometimes on a mass scale, for instance at the Allen Brain Institute in Seattle. However, for the moment, researchers are predominantly using the somewhat smaller, more fathomable and available cerebral matter of fruit flies and mice. As neurobiologist Richard Wingate and many of his colleagues freely admit, the human brain remains a largely mysterious world, and its explorers, surveyors and cartographers have only begun to identify its main roads, rivers and regions, albeit in spectacular and prolific ways.

"the human brain remains a largely mysterious world, and its explorers, surveyors and cartographers have only begun to identify its main roads, rivers and regions, albeit in spectacular and prolific ways."

A political economy of brains

Neuroscience may, in the words of critical insider Steven Rose, be "data-heavy, and theory-light", but in the USA especially it enjoys a prestige which derives partly from its stance as an objective material science. Some would regard it as the tech-savvy neoliberal nephew of Cold War psychology. Any trawl of an airport bookshop will reveal the promises of applied neuroscience to make you a better manager, marketer, citizen, parent, law-enforcer and consumer. With the prospect of manipulation through smart drugs and molecular interventions, a whole neurotechnological civilisation is beckoning. It is easier, as Hollywood realised, to blame or speculate about the brain than to change the social and economic conditions that many would still argue are crucial factors in mental illnesses such as schizophrenia, depression and post-traumatic stress disorder. Some of the claims made for neuroscience by the mainstream media are scarcely less fantastic than the dystopian visions conjured up by a recent sub-genre of neuro-fiction movies, such as *The Matrix* (1999), *Inception* (2010) and *Limitless* (2011). In all seriousness, US judges are now reported to be considering whether information held within a defendant's brain can legally be considered a physical thing and therefore, once extracted by fair means or foul, admissible as evidence in court cases. This would in effect remove the protection

**Brain Tissue Freezer 1
(from the series 'After I'm Gone')**
Ania Dabrowska, 2011
C-print
COURTESY OF ANIA DABROWSKA AND BRONWYN PARRY

of the Fifth Amendment, which prevents defendants from being forced to testify against themselves. The brain, under the new, penetrative regime of neuroscience, risks being exiled from its longstanding domain of the self.

Whether or not the brain remains the sovereign organ, it is, like any other substance or thing, part of a system of production, circulation and exchange, as the economic metaphors surrounding it suggest. Scientists 'harvest' and 'bank' brains in order to produce value in the form of scientific knowledge, professional advancement and public health; teachers 'cultivate' intelligent minds. We have seen how popular culture, through horror and science fiction, also perpetuates the idea that brain tissue is a thing of extraordinary power and potential, the core of the perennial trophy of the severed head. These complex processes have visible and tangible effects. Brains are displayed, pictured, scanned, analysed and treated, thus transforming them by scientific industry into academic papers, media spectacles of discovery, education and information, healthier citizens, more compliant workers and consumers – all to enhance national prowess, prosperity, security and so on. A multitude of spaces, aesthetic predilections, protocols, technologies and tools supports and enables this range of practices. There is, so to speak, a political economy of brains, by which the brain becomes invested with a level of hope and expectation that has usually exceeded the ability of science to make sense or use of it. The brain has, therefore, in the classical sense, considerable commodity value, as the Californian conceptual artist Jonathon Keats satirically pointed out when he sold futures contracts in his brain for $10 per million neurons.

> "There is, so to speak, a political economy of brains, by which the brain becomes invested with a level of hope and expectation that has usually exceeded the ability of science to make sense or use of it."

While this book also capitalises on brains, it does so with their original owners firmly in mind. The images and objects arranged here are traces of human behaviours and experiences, rather than evidence towards some theory of mind or body (though they may have at one time served that purpose). Some of these episodes were cruel and disastrous, others noble and optimistic; some were unremarkable by the standards of the time but would be unacceptable today. All the exhibits are chosen because they consist of, directly involve or somehow represent the form and matter of the brain. They indicate how brain tissue has served as cultural material: a natural resource that humans have worked to produce meaning and perceived individual or social benefit, as well as profit, mainly in medical or scientific contexts. The emphasis is therefore on collecting and other direct physical interactions with brains, not chemical interventions with drugs, which are nevertheless important and have biophysics at their root. There are also a number of reflections by contemporary artists on these processes, some of them specially commissioned for this project.

The selection of cases is neither exhaustive nor encyclopedic, but designed to highlight some of the most potent stories that emerge from past and present neurology. It is based partly on the great collection of historical and ethnographic artefacts and books amassed by Sir Henry Wellcome (1853–1936), the library and some objects from which are housed at Wellcome Collection in London. Though that collection covers an enormous cultural and historical range, the present selection is mainly from Europe and North America from the 16th century onwards. It is divided into four sections, reflecting impressionistic themes that emerge from the encounter with the brain as a thing. The first, Measuring/Classifying, explores the anthropometric motivations that lay behind much brain science. The second, Mapping/Modelling, shows some key attempts to represent the anatomy of the brain. The third, Cutting/Treating, considers surgical and anatomical interventions. And the fourth, Giving/Taking, explores some of the exigencies and salient narratives of brain-collecting to the present day. It is hoped that together, they show the immense weight of hopes, fears and beliefs with which this delicate organ has been saddled, and so open up some productive questions about our collective minds.

Examining the brain
Richard Wingate

Whether or not the human brain is the most complex object in the universe, it is undoubtedly something that all of us find fascinating: our brains are taken with brains. From the spoken word or gesture to the flicker of emotion in a fleeting expression, an enormous portion of the brain's workings is focused on analysing the external manifestations of processes taking place inside other people's heads. Evolution has honed both our ability to display an inner world of thoughts and emotions through speech and movement, and our sensitivity to the intended and unintended messages similarly conveyed by others. This semaphore of sound and body is driven and decoded by the one to two kilos of jelly-like biological machinery concealed within our bony cranium. Since the early 1800s, we have increasingly realised that everything that we think, feel and plan – the totality of tasks our mind performs – is registered in material changes within the brain.

It is ironic, then, that there is so little to read of ourselves in the material stuff of the brain. Despite all that we intuitively understand of the rich complexity of the mind, the brain itself gives little if anything away. As an isolated object, its inner workings are completely inscrutable. Cold to the touch and superficially bloodless, it seemed to Aristotle like a cold sponge; he proposed that its function was to mop up the heat generated by the thinking, feeling heart. The brain has no obvious moving parts, no pumps, valves or tendons. Its pattern of corrugations, swellings and folds has no immediate reference to any easy meaning. Attempts to gauge moral capacity and IQ in terms of the size, shape, symmetry and topography of our brains have proved fruitless. Any trace of personality, gender or intelligence is absent. Great men and women do not possess great brains, or at least their greatness cannot be registered in the size or shape of their brains. Unless afflicted by a grievous disease or injury, its structure tells us nothing about the life of its owner. Nevertheless, as it is the place in the body where the self resides, we have become accustomed to imagining changes in our psyche in terms of physical changes to synapses, connections and nerve cell numbers. We fear for its structural integrity, its susceptibility to drugs and alcohol, and its legendary frailty in the face of bumps and sneezes. We are all too conscious of its notorious inability to regenerate. And, ultimately, even though we have a wide selection of bodily organs that are of critical importance to our survival, it is by the brain's death that we measure our own.

Architectonic fields of the cerebral cortex (below)
Korbinian Brodmann, 1909
Illustration in book

KORBINIAN BRODMANN MUSEUM, HOHENFELS, GERMANY

Fixed 2 (from the series 'After I'm Gone') (bottom)
Ania Dabrowska, 2011
C-print

COURTESY OF ANIA DABROWSKA AND BRONWYN PARRY

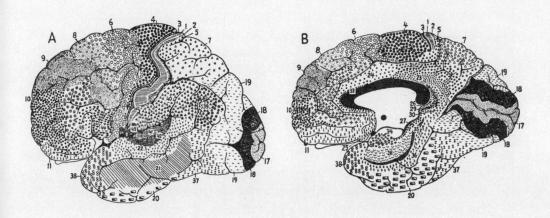

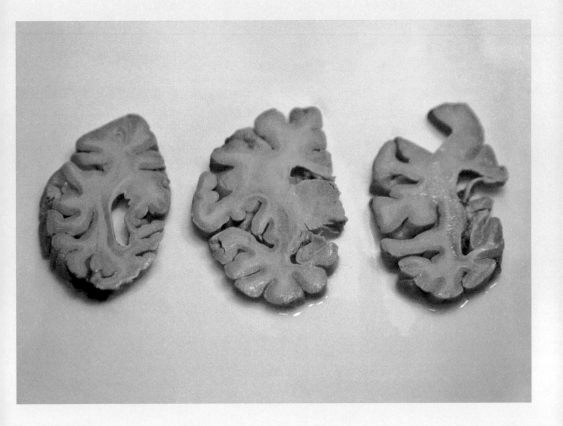

Digging deeper

If surface anatomy cannot show how our concept of self relates to the materiality of the brain, does a slice into the brain give any answers about the origins of individuality and intellect? A section through the brain's structures reveals patterns that are tantalising, but these proved every bit as indecipherable to the pioneer anatomists of the 17th century – the early Enlightenment – and onwards. The nomenclature of colour and texture that is so often used to describe this territory speaks of an anatomical exasperation at the lack of functional clues. The genu is knee-shaped, the striatum is striated and the corona radiata radiates outwards, while the substantia nigra, locus coeruleus and nucleus ruber are black, blue and red, respectively. For all that their shape and spatial organisation hint at subterranean pathways and hubs for the interchange of the mental traffic, the internal structure of a slice is as unyielding to interpretation as the surface of the brain.

> "As an isolated object, the brain's inner workings are completely inscrutable. Cold to the touch and superficially bloodless, it seemed to Aristotle like a cold sponge; he proposed that its function was to mop up the heat generated by the thinking, feeling heart."

This lack of meaning in a 'reading' of neuroanatomy was compounded by early forays into microscopy that revealed the fabric of the brain to be a densely tangled meshwork of threadlike tendrils. The contours of this impenetrable forest suggested highways and byways of information flow, yet optical and histological (slide-making) techniques – with which Robert Hooke and Antonie van Leeuwenhoek had so successfully revealed the fundamental, 'cellular' organisation of organisms in the 17th century – failed to decipher the brain's internal structure for another 200 years. Indeed, by this point, the microscopic fabric of the brain was seen as an exception to the general cell theory, as a sea of fibres and tubes that conducted thoughts seamlessly around the central nervous system.

What unlocked the microanatomy of the nervous system was an accidental discovery in 1873, by the Italian scientist Camillo Golgi, of a chemical reaction that darkly stained only a fraction of the brain's material substance. This technical breakthrough was driven by a conjunction of improvements in the optical quality of the microscope objective lens, the introduction of organic preserving fixatives from a newly industrialised Germany, and experiments with early photographic solutions. The partial, fragmentary view offered by Golgi's 'black reaction' afforded a crucial insight. By piecing together the strands that appeared from out of a forest of complexity, Golgi

The cells of the cerebellum
Camillo Golgi, Calderini, Reggio Emilia, 1885
Illustration in book
UNIVERSITY HISTORY MUSEUM OF THE UNIVERSITY OF PAVIA

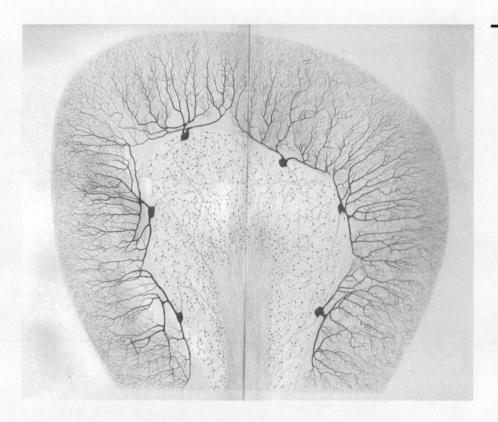

(and his great rival, the Spanish anatomist Santiago Ramón y Cajal) meticulously reconstructed the cells of the brain. But, in a story that has become the stuff of scientific folklore, Golgi failed to recognise that the tendrils he had discovered were parts of brain cells, and he persisted with a non-cellular theory of brain structure that condemned his beautifully drawn observations to relative obscurity. By contrast, it was Cajal who intuitively used the products of Golgi's reaction to reconstruct a tapestry of extraordinarily exotic branching forms of neurons in a distinctly (if not extravagantly) cellular fabric.

Cajal was captivated by the possibilities of Golgi's method and in the 1880s commenced a monumental mapping project. He documented the microscopic landscape lying beneath the laminae, nuclei, gyri, sulci, pyramids, olives and ganglia of the nervous system, revealing a rich pattern of regional architecture. Major neuroanatomical structures, such as the cerebellum and cerebral cortex, could be seen to contain microscopic organisational motifs that not only transcended individual differences but also were remarkably constant across species. Here was a logic that allowed

Parasagittal section of the cerebellum
Santiago Ramón y Cajal, 1894
Ink drawing on paper
COURTESY OF CAJAL LEGACY, INSTITUTO CAJAL (CSIC), MADRID

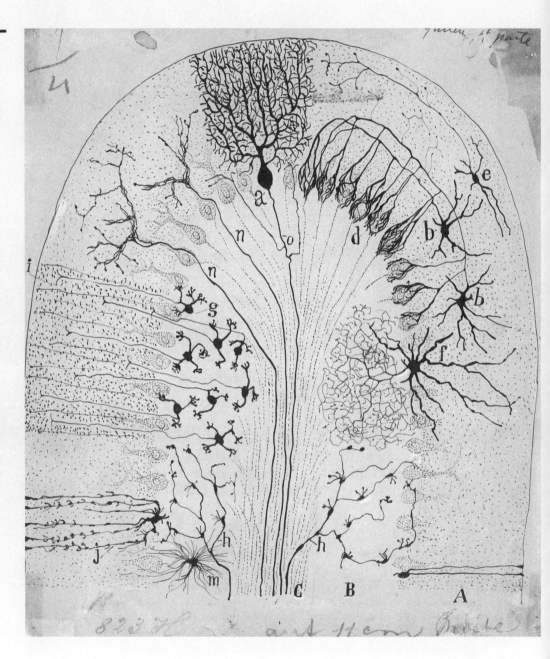

Cajal to propose general laws of information flow that still hold today. In contrast to the ambiguity of the brain's external structure, the microscopic landscape seemed to speak clearly of its function. More than simply showing the brain's folds and recesses with greater clarity, the discovery of its cells revealed an entirely new level of orderly microarchitecture for the machinery of thought, which could only be glimpsed in fragments at the limits of technology.

Seeing the brain in a new light
Since the discovery of the brain cell by Golgi and Cajal, and as technology has become ever more ingenious, the significance of the material, touchable stuff of the brain has become ever more eroded. At the microscopic level, we can now use powerful pulsed laser illumination to inspect the function of synapses in real time, deep within the intact brain. By coupling the genetics of biofluoresence with powerful three-dimensional imaging, the dynamics of interactions between specific proteins, or the transport of ions into the branches of neurons, can be mapped and isolated from a teeming sea of activity. At the macroscopic scale, tomography using a variety of different electromagnetic wavelengths has re-designated the outline of the brain as a frame for displaying the patterns of oxygen usage, water content or radiation absorbency in populations of cells. Brain material is re-engineered as a template for the display of blobs of colour in a thermal digital palette of red hotspots and blue lacunae. In the latest developments of magnetic resonance, diffusion tensor imaging can reveal a spectacular framework of connectivity by measuring the polarity of water molecules constrained by the fine strands of brain-cell axons.

"More than simply showing the brain's folds and recesses with greater clarity, the discovery of its cells revealed an entirely new level of orderly microarchitecture for the machinery of thought."

These digital representations have become the clinical focus for those who look after and fix our damaged brains. The physical material of the brain is at best inaccessible, and most clinicians are more familiar with an optical section through the central nervous system than a physical slice of post-mortem material. Computerised reconstruction produces a facsimile of material substance that can be rotated, re-rendered, paused or magnified. For the scientists who investigate the function and structure of individual brain cells, the digitally realised view of the brain's structures is the arena in which ideas and theories are conceived and developed. And culturally, who would doubt that these techniques have also generated a more appealing and even aspirational vision of our brain? In place

Brainbow Mouse
Jean Livet, Joshua R Sanes and Jeff Lichtman, 2007
Digital photomicrograph
LICHTMAN LAB, HARVARD UNIVERSITY

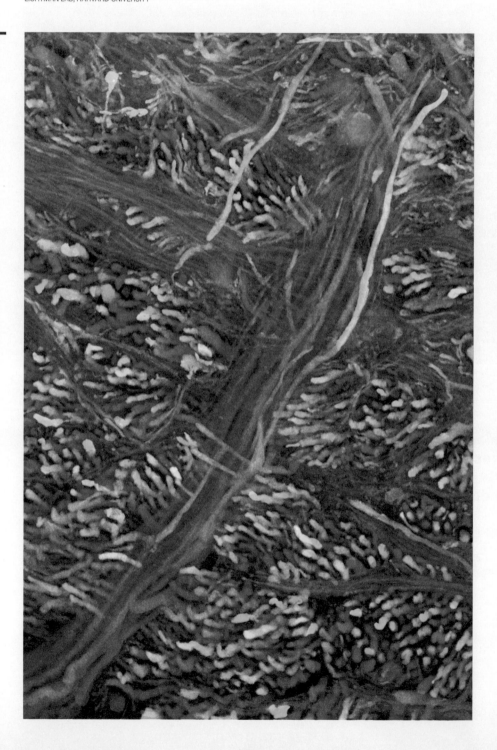

of an alien, embalmed, grey, strangely textured and intractable anatomised object, technology offers a spectacular digital form of infinite complexity, displayed in aesthetics of biofluorescence. What started with Golgi's unexpected discovery has evolved into a virtual transcendence of the corporeal brain from Aristotle's 'cold sponge' into a visually ethereal realm of light and colour.

Getting to grips with authenticity

Given all this, why then do we anatomy teachers spend hours exhorting students to study the surface features of isolated brains, the torn and hard dissected remnants of its blood supply and, in brain slices, the subtle textures and colours of an older school of anatomy? Why is the material of the brain still powerful enough to command our attention?

> "In place of an alien, embalmed, grey, strangely textured and intractable anatomised object, technology offers a spectacular digital form of infinite complexity, displayed in aesthetics of biofluorescence."

Part of the answer is that the real brain fills space and has tangible volume. While the digital brain is stretched and shrunk to fit the page or screen, our heads are containers for nervous substance, snugly encased in membranes and bone. At this tight fit between bone, brain, membrane and blood vessels lie faultlines that generate aneurysms, strokes and intracranial bleeds. The physical vulnerability of these structures is there to be seen in the totality of anatomy, but can sometimes be lost in the selective vision of the digital image. However, there is also a sense of theatre in dissection, and presence in the prosected artefacts that it produces. What the real brain embodies is no less than a sense of its own authenticity. The privilege of observing the inner world of the real, anatomised body was historically granted to few and today is strictly regulated and safeguarded. To touch the human brain is rare. While the more accessible digital world is synthetic and detached from the material substance it portrays, the reality of human material used in the anatomy class is never in doubt. This can perhaps be likened to the contrast between handling a priceless museum artefact and a faithful replica. For both brain and artefact, authenticity comes from provenance, and although anatomy donors are scrupulously kept anonymous, the undeniable origin of a real brain gives it a presence that commands a different kind of attention than a model can.

Authenticity is significant. It seems to prompt different ways of handling and describing material. The topography of surface and slice becomes a signpost to the inner microscopic fabric of nerve

**Brain tractography:
visualising neuronal tracts**
Diffusion tensor image, 2011
NUADA MEDICAL/WELLCOME LIBRARY, LONDON

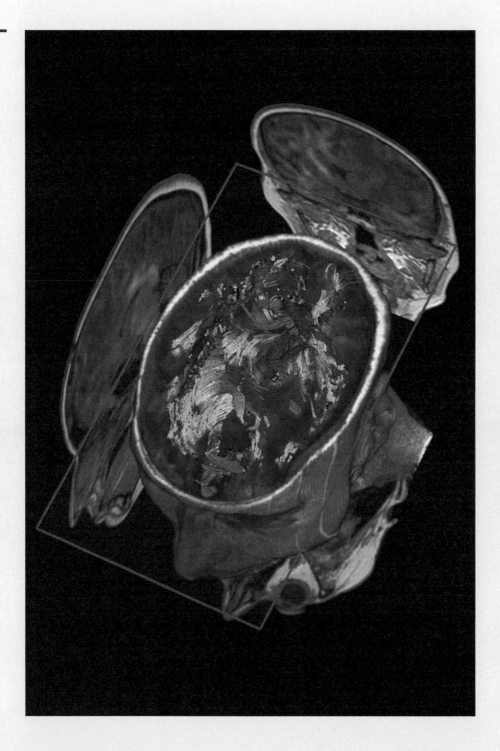

cells. The names of landmarks recall the colonisation of neural territory by eponymous 19th- and 20th-century anatomists: Broca, Golgi, Clarke, Russel, Cajal, Wernicke, Purkinje and Deiters all gave their names to structures they encountered. In the digital image, the process of visualisation is integral to the meaning of what is being seen, while in the unmediated brain the different functions lie concealed beneath seemingly unrelated grooves and hillocks. As we weigh a brain in our hands and trace its contours with our fingers, we can physically touch the centres for speech, movement, hearing and vision in different lobes, and we can trace the course of the electrical stimulations that the Canadian neurosurgeon Wilder Penfield used to map the sensations from the body onto the surface of the cerebral cortex.

> "As we weigh a brain in our hands and trace its contours with our fingers, we can physically touch the centres for speech, movement, hearing and vision in different lobes"

This material also speaks of something more, beyond a tapestry of the historical, cultural or functional: something that again cannot be readily found in the selective and crystal-clear digital incarnations of the brain. Between the landmarks, and in the infinitesimal detail of a vastly branching cellular panorama, are places where knowledge falters and only questions remain. While we have found out much about how the brain works, the yawning gaps in our understanding are reflected in territories where description is sparse. What we are going to find here is uncertain: fractal parcelling of function into smaller and smaller definable territories? The store for a single memory? Or perhaps the coordinates of consciousness itself? Just like the medieval atlas, our physical brain speaks of large cartographic blanks, foreign lands that have rarely been visited and are populated by strange and exotic creatures. Here be dragons!

Photographic essay: Brain collecting
Daniel Alexander, 2011
C-prints

The historic Charité hospital in Berlin was founded in the early 18th century as a plague quarantine, a workhouse for beggars and a maternity hospital for unmarried mothers. It was later developed as a military and teaching hospital. Many of its present red-brick buildings date from the Wilhelminian period before World War I, and sustained substantial damage during World War II.

The anatomy collection was assembled by the famous pathologist Rudolph Virchow (1821–1902), who argued that studying the development of cells was essential to the understanding of disease. He made his museum available not only to students and the medical profession but also to the public. For many years, the brains not on public display were stored in the old chapel that adjoined the dissection rooms and anatomy theatre. In July 2011, with the chapel due for renovation, they were moved along with other anatomical preparations to the arched cellars underneath these buildings, leaving the empty shelves labelled 'Gehirn' ('brains').

Both the old and the new quarters for the brains were recently photographed by Daniel Alexander as part of a special commission for this project. Taken on large-format colour film, these pictures document not only a mode of collecting brains, but also the brains' changing architectural settings.

Near the Medical Historical Museum

The north-east corner of the Campus Charité Mitte, Berlin. Left to right: autopsy rooms and pathology labs, chapel, anatomy theatre.

Chapel west door

Anatomy specimens are stored in the cellar.

Chapel north end

The former brain stores now awaiting conversion. These recently vacated cabinets held the specimens; the motto around the rose window reads 'Christ Is My Life, Death Is My Victory'.

Medical Historical Museum cellars 1
(above; detail, pages 36–37)

From 2011, the new resting-place for the brains.

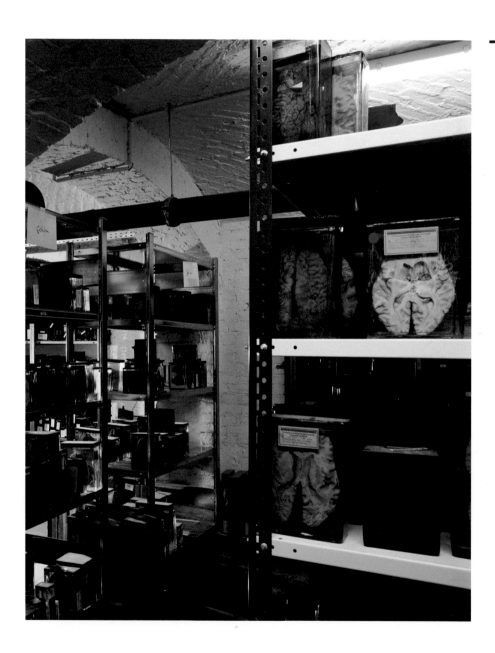

Medical Historical Museum cellars 2

The specimen on the right marked with a red dot is the brain of a woman who was shot in the head aged 19 but survived until late middle age.

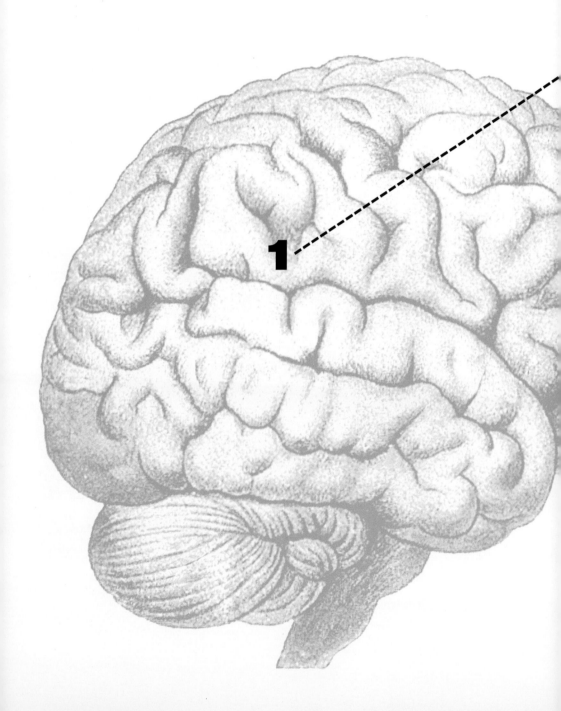

Chapter One

Measuring / Classifying

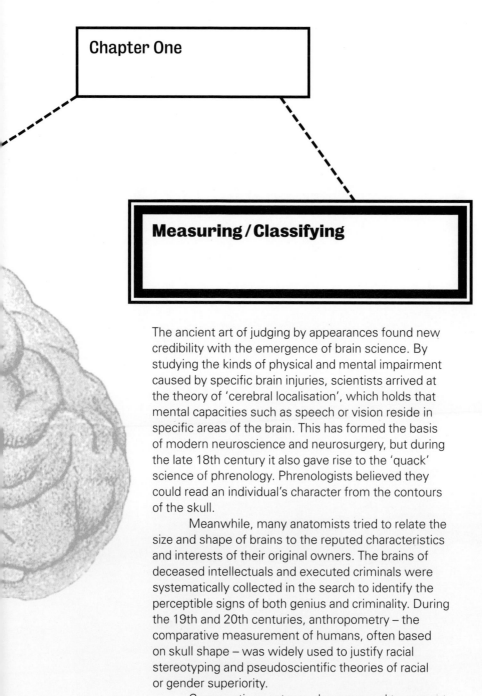

The ancient art of judging by appearances found new credibility with the emergence of brain science. By studying the kinds of physical and mental impairment caused by specific brain injuries, scientists arrived at the theory of 'cerebral localisation', which holds that mental capacities such as speech or vision reside in specific areas of the brain. This has formed the basis of modern neuroscience and neurosurgery, but during the late 18th century it also gave rise to the 'quack' science of phrenology. Phrenologists believed they could read an individual's character from the contours of the skull.

Meanwhile, many anatomists tried to relate the size and shape of brains to the reputed characteristics and interests of their original owners. The brains of deceased intellectuals and executed criminals were systematically collected in the search to identify the perceptible signs of both genius and criminality. During the 19th and 20th centuries, anthropometry – the comparative measurement of humans, often based on skull shape – was widely used to justify racial stereotyping and pseudoscientific theories of racial or gender superiority.

Comparative anatomy also appeared to support the view that big brains distinguished *Homo sapiens* from other species and from our prehistoric forebears. However, most modern neuroscientists would dispute any simple correlation between brain mass and social attributes such as intelligence.

Headspanner in a wooden box (below)
c.1896
Wood

This is a device for measuring the size of human heads (and hence brains), developed by Sir Francis Galton (1822–1911). Known as the 'father of eugenics', Galton was influenced by the evolutionary theories of his cousin Charles Darwin and believed in the need to improve human races via selective breeding. Using a variety of 'anthropometric' devices to measure the proportions of the human frame, he sought evidence of links between physical appearance and the supposed evolutionary progress of different population groups.

THE GALTON COLLECTION, UNIVERSITY COLLEGE LONDON

Brain of a female common dolphin (right)
19th century
Wet specimen

Dolphins have a long history of interaction with humans and are renowned for their intelligence. They show a remarkable ability to learn, communicate, and form concepts and cultures. Some theorists relate mammalian intelligence to numbers of cortical neurons. Bottlenose dolphins are estimated to possess just over half as many of these (5.8 billion) as humans – between chimpanzees and gorillas, but fewer than elephants.

HUNTERIAN MUSEUM, ROYAL COLLEGE OF SURGEONS, LONDON

MEASURING/CLASSIFYING 43

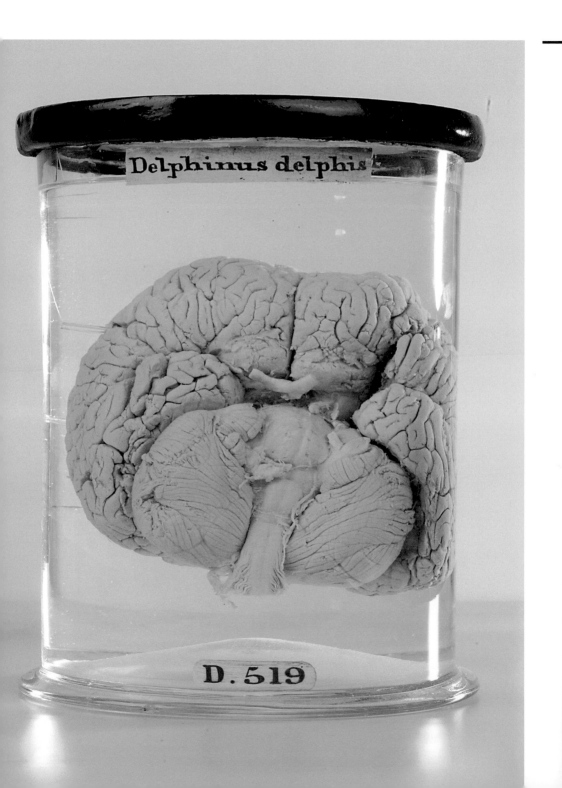

Alfred Haddon measuring heads in the Aran Islands
1892
Photographic lantern slide

The Aran Islands off the west coast of Ireland were romanticised by Victorian folklorists, antiquarians and novelists as the last bastion of a 'pure' Irish people and its 'primitive' culture. Alfred Cort Haddon (1855–1940) was one of the founding fathers of British anthropology and co-founded the Dublin Anthropometric Laboratory in 1891. With the assistance of Charles Browne and the police, he measured 27 men, mostly on the island of Inis Mór. "On the whole," they concluded, "the people are decidedly good-looking."
MUSEUM OF ARCHAEOLOGY AND ANTHROPOLOGY, UNIVERSITY OF CAMBRIDGE

Teaching measuring techniques in India
c.1850–95
Photographic lantern slides

The magic lantern projector proved a valuable means of instruction as well as entertainment from the 17th to the 20th centuries. These two images taken from a 19th-century British textbook on Indian administration suggest the crucial role of colonial authorities in gathering anthropometric data from often resistant populations. Such practices were widespread among the European empires. Some methods, such as the making of plaster casts from faces, were brutally applied and sometimes resulted in the deaths of subjects.

MUSEUM OF ARCHAEOLOGY AND ANTHROPOLOGY, UNIVERSITY OF CAMBRIDGE

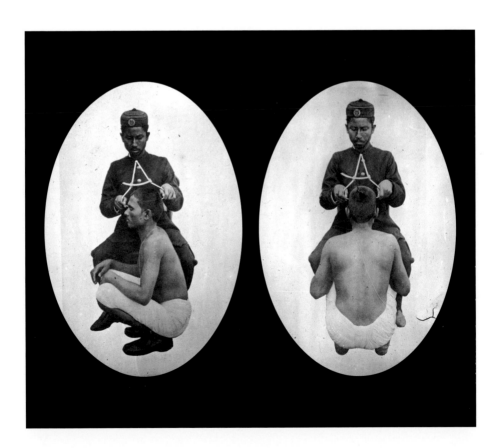

Cast of microcephalic skull (Victoire) (below)
*c.*1830
Plaster copy

This is a plaster copy of a cast from the skull of a young woman named Victoire, described as an idiot and suffering from microcephaly, a rare genetic condition that restricts brain development. In general, individuals with microcephaly suffer impaired brain function and reduced life expectancy. The incision around the cranium indicates that Victoire's brain was removed for post-mortem study, and probably also cast.

SCIENCE MUSEUM, LONDON

Phrenological head (Victoire) (right)
*c.*1825
Plaster copy

This cast accompanied Victoire's skull in the collection of the British Phrenological Society, which also included plaster casts of the heads of eminent persons and the skulls of apes for comparison. Phrenology was controversial and dismissed by many as pseudoscience, but the British Phrenological Society survived until 1967.

SCIENCE MUSEUM, LONDON

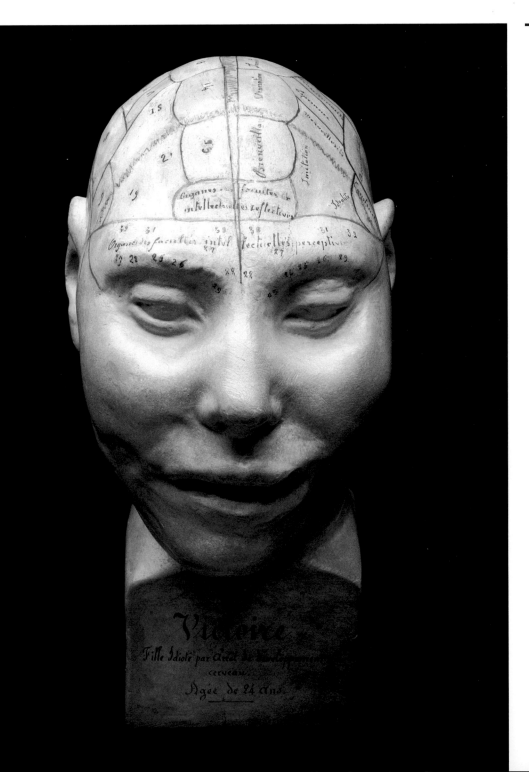

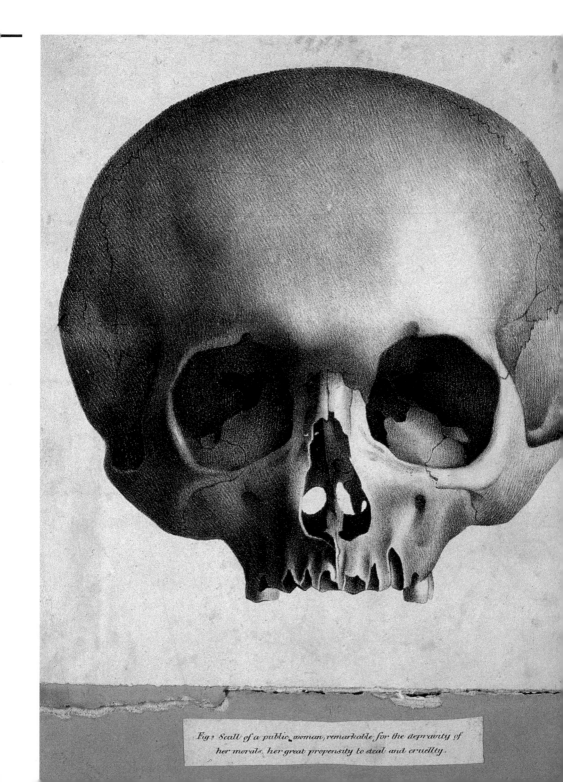

Fig. 1 Scull of a public woman, remarkable for the depravity of her morals, her great propensity to steal and cruelty.

MEASURING/CLASSIFYING 49

Skull of a prostitute who accompanied the army and was reputed to be cruel and violent, frontal view (left)
Bernard Hollander after Joseph Vimont, early 20th century
Lithograph and text mounted on board

Bernard Hollander (1864–1934) was a London-based psychiatrist from Vienna who attempted to revive phrenology as a respectable science. This illustration was removed by Hollander from Joseph Vimont's *Traité de Phrenologie Humaine et Comparée* (1832), and was intended to demonstrate the theory that criminal tendencies can be related to the shape and size of the skull. This particular skull is thought to have been from the collection of Franz Joseph Gall (1758–1828), founder of phrenology.
WELLCOME LIBRARY, LONDON

Skull of the painter Raphael (below)
Bernard Hollander after Joseph Vimont, early 20th century
Lithograph and text mounted on board

Also extracted by Hollander from Vimont's *Traité de Phrenologie Humaine et Comparée*, the example of Raphael's skull implies that superior as well as inferior intelligence could be clearly identified by observations of the head. Vimont (1795–1857) was a comparative phrenologist whose book on the subject contained many images that had been borrowed from Gall's *The Anatomy and Physiology of the Nervous System in General, and of the Brain in Particular* (1819).
WELLCOME LIBRARY, LONDON

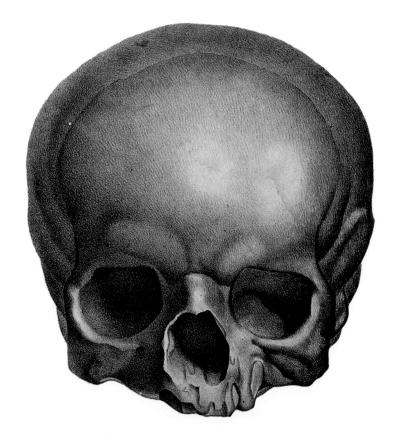

Bernard Hollander illustrating his system of cranial measurement (below)
Bernard Hollander, c.1902
Photographic contact sheet (detail)

Bernard Hollander introduced a methodology for measuring the skull and comparing the results with statistical averages. He was also concerned with studying localisation in the brain of the fundamental psychical functions. Hollander's system explicitly sets aside a faculty for 'religious sentiments', whereas older phrenological systems tended to separate these into 'veneration', 'sublimity' and 'hope'. However, like Franz Joseph Gall, Hollander associates the cerebellum at the base of the brain with the libido.
WELLCOME LIBRARY, LONDON

The Brain of the American Negro (right)
Scientific paper by the American anatomist and brain-collector Burt Green Wilder (1841–1925), 1909

After graduating from Harvard, Wilder, who was white, served as a surgeon in the Fifty-fifth Massachusetts Infantry, a black regiment in the American Civil War. Inspired by his comrades, he became an advocate for African-American civil rights. He read this paper to the First National Negro Conference in New York in 1909. Though Wilder's methodology may seem offensive today, many of the conference resolutions – equal opportunities and spending on state schools – still resonate.
DIVISION OF RARE AND MANUSCRIPT COLLECTIONS, CORNELL UNIVERSITY LIBRARY

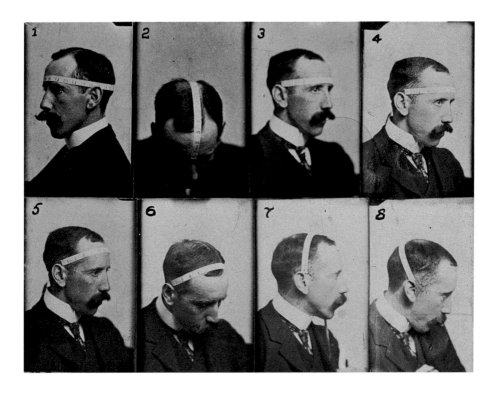

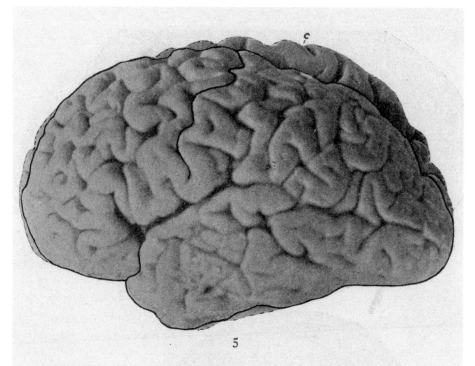

Fig. 5. The continuous black line is the outline of the left cerebral hemisphere of No. 3652; the interrupted line is the outline of the right hemisphere of No. 2912. They were photographed together so as to be about one-third natural size. On 2912 the Central fissure is marked C; on 3652 its course is shown by the undulating black line; the shorter line at the lower margin represents part of the Sylvian fissure.

These are the opposite halves of the cerebrums of two very unlike persons. The right half is from G. F., an illiterate black janitor. The left from a white jurist and politician. As an ally of Tammany Hall he probably condoned, if he did not encourage, the race riots in this city in the spring of 1863 when the first northern colored troops enlisted in spite of Democratic opposition. If so, we may charitably ascribe his conduct to sharing the general belief that every Negro's brain is so small as to unfit him for citizenship or even for military service. Yet the brain of the black janitor weighed 5 ounces more than that of the white jurist (Table 3), and now, when the left half of the latter is held against the right half of the former so that the lower margins coincide, at nearly all other points the black's outline may be seen beyond the white's. Let us hope that X. Y. Z. now rejoices that at least one of the blunders of his life has been rectified after his death.

52 BRAINS: THE MIND AS MATTER

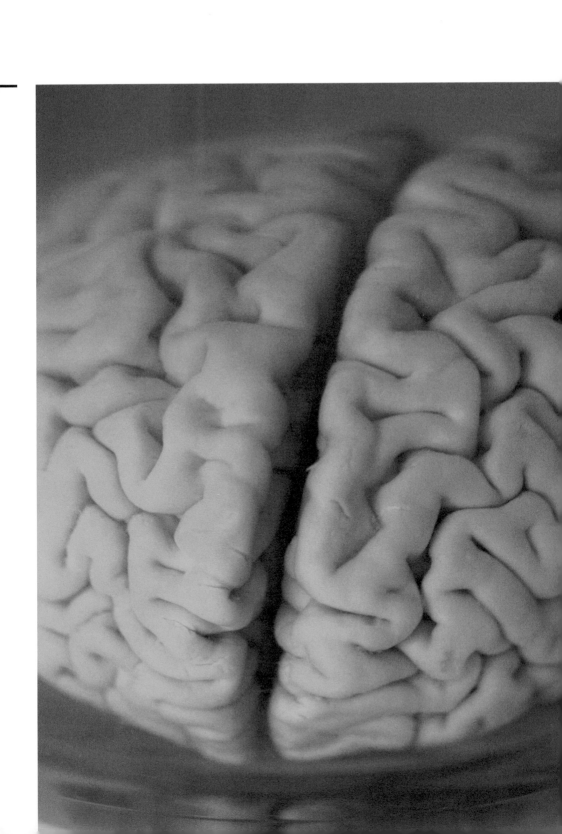

Preserved brain of Helen H Gardner, women's suffragist
d. 1925
Wet specimen

The brain of American writer, civil servant and feminist Helen H Gardner (b. 1853) is one of 70 still remaining in the Wilder Brain Collection at the Department of Psychology, Cornell University. She argued that the female brain "was not demonstrably different from that of a man under the same conditions and with the same opportunities for development."
DEPARTMENT OF PSYCHOLOGY, CORNELL UNIVERSITY

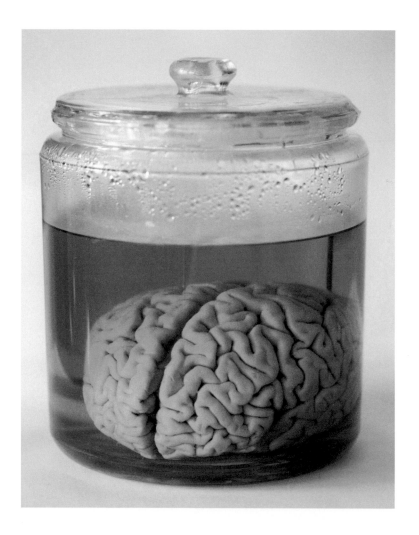

Preserved brain of Edward H Rulloff, "gentleman, scholar and murderer"
d. 1871
Wet specimen

The brain of Edward H Rulloff (b. 1819/20), is perhaps the most infamous in US history. Self-educated, Rulloff excelled in languages and financed his researches through crime sprees, including the probable murder of his wife, son and two accomplices. He went to the gallows for killing a shop assistant in Binghampton, New York. Mark Twain wrote a satire suggesting that another man be hanged in his place to spare his astonishing intellectual gifts. Rulloff's brain weighs 1770 grams – the biggest in Wilder's collection and one of the heaviest preserved brains extant.

DEPARTMENT OF PSYCHOLOGY, CORNELL UNIVERSITY

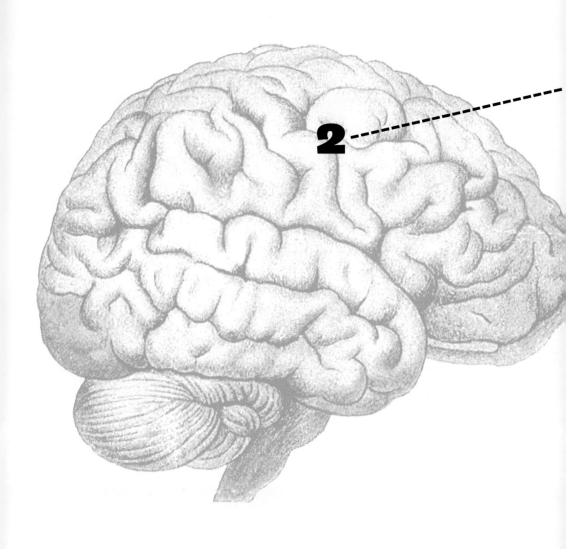

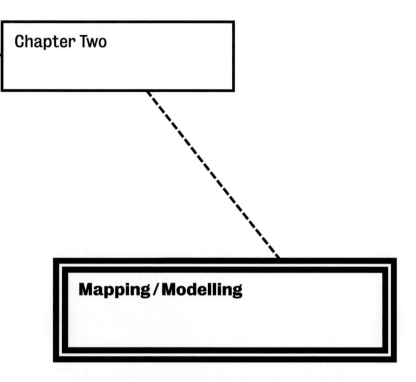

Chapter Two

Mapping / Modelling

Representing the complexity of the brain has long been a challenge for medical artists. Coloured wax has been used since ancient times to simulate flesh, and during the 19th century European model-makers became world-renowned for their skill and accuracy in wax and other materials such as plaster, papier-mâché and wood. However, some models, if they were only needed to help students to memorise the main features of the brain, were ruthlessly simple.

In the graphic depiction of the brain, developments from the Renaissance onwards tended towards greater visual objectivity and the exclusion of contextual features. In the late 19th century, new microscopic staining techniques enabled the discovery of the tree-like neuron, the brain cell. This revolutionised our understanding of brain structure and function, and offered appealing glimpses of what looked like a deep forest with innumerable connections between branches.

Though computer scanning and modelling now provide exquisitely detailed and versatile images of the brain, physical models made from plastic and other materials still play an important part in training doctors and surgeons and conveying concepts and information.

Box model of the brain
Mid-20th century
Photographic prints mounted on wood, paint

This unique and somewhat ad hoc model does not pretend to resemble the actual shape of the brain, yet provides a surprisingly effective tangible and three-dimensional representation. It was used in anatomical teaching in the mid-20th century to show the principal areas of the brain. Believed to have been made in the anatomy department at the University of Aberdeen, it appears to have been well used by students, who have even added their own labels.
ANATOMY MUSEUM, UNIVERSITY OF ABERDEEN

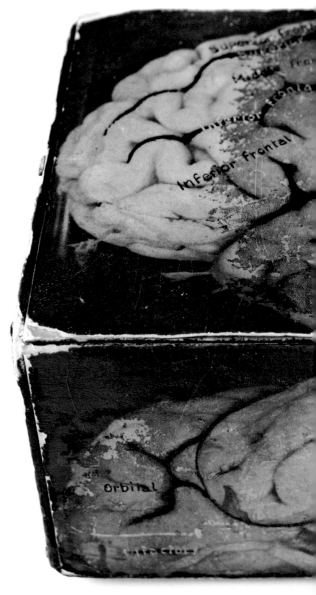

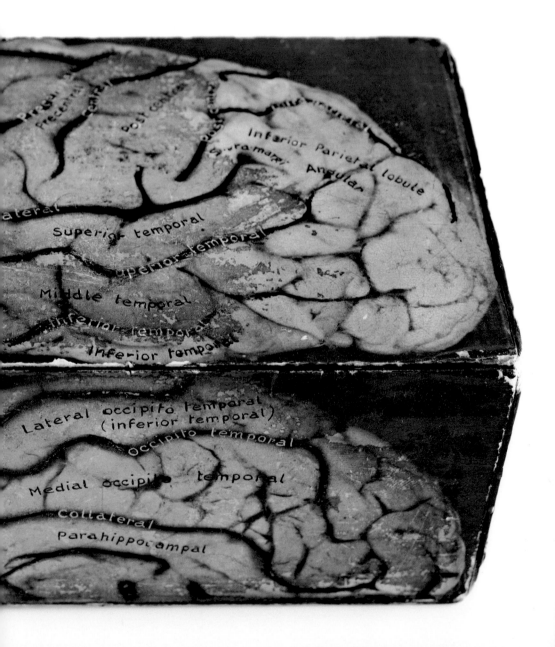

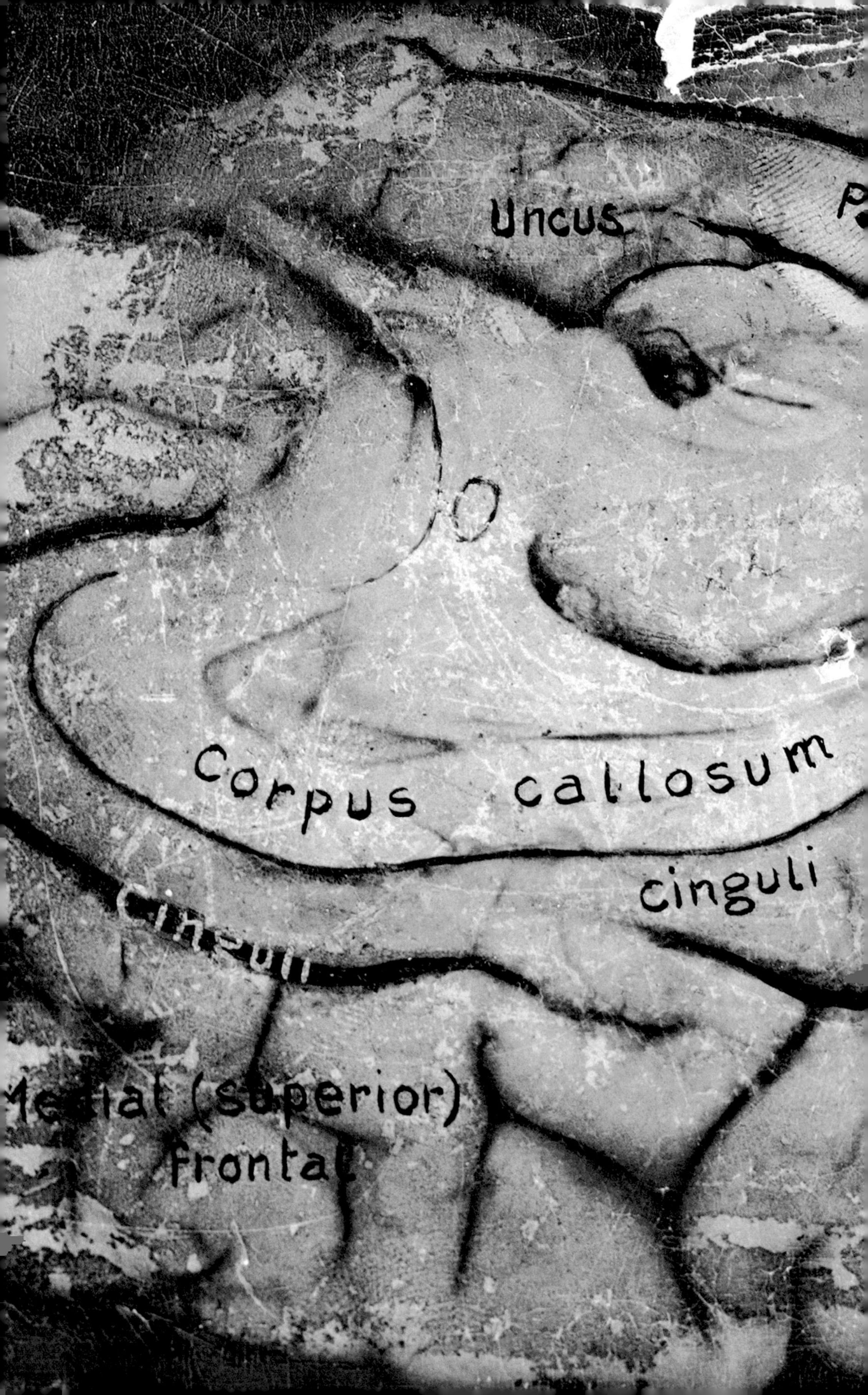

Uncus

Corpus callosum

cinguli

Cinguli

Medial (superior) frontal

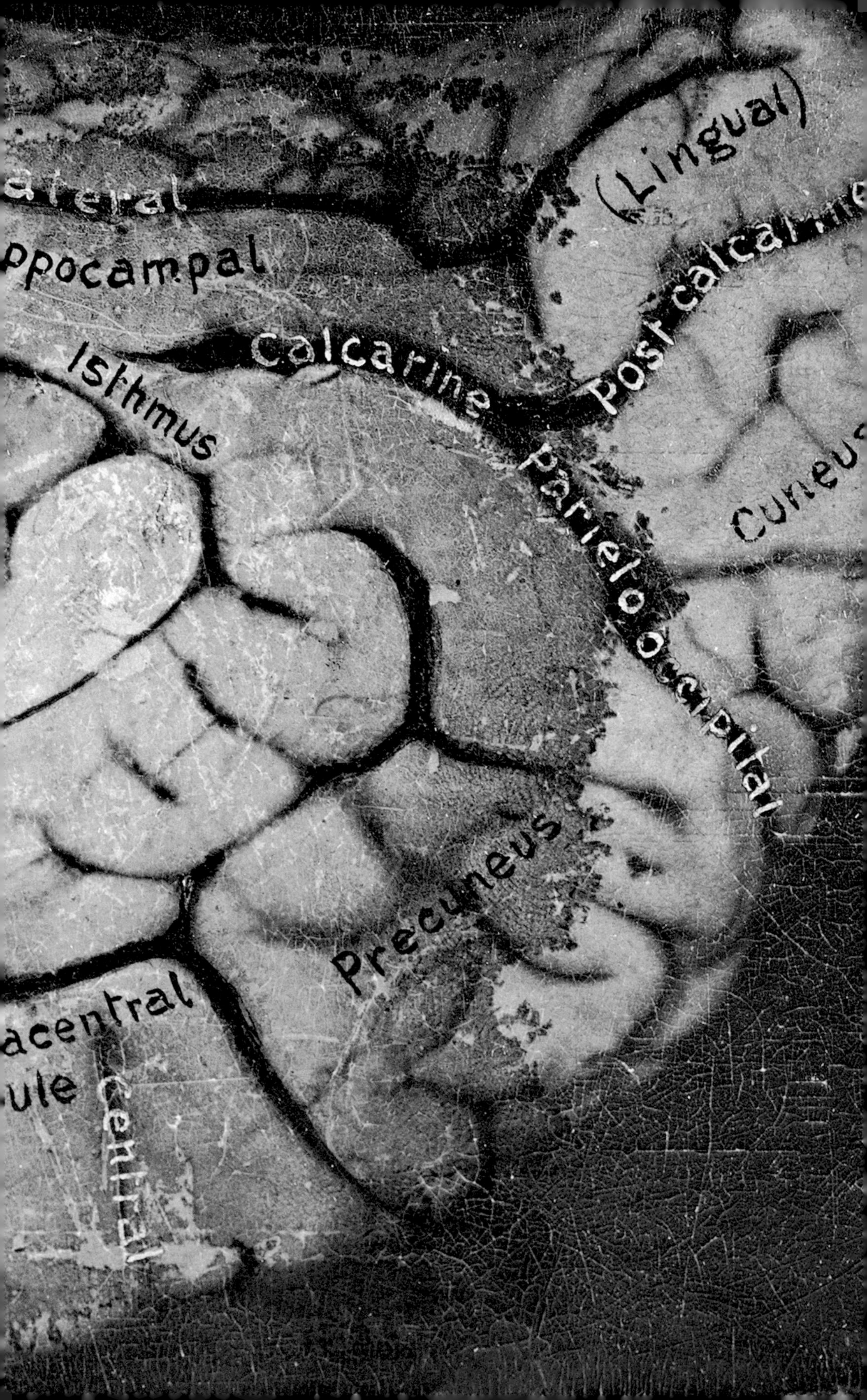

...ateral
...ppocampal
(Lingual)
Isthmus
calcarine
Post calcarine
Parieto-occipital
Cuneus
Precuneus
...acentral
...ule
Central

Diagram of the Soul (below)
Margarita Philosophica nova
Gregor Reisch, 1508
Woodcut illustration in book

In the medieval period, the localisation of mental processes was considered to be in the ventricles or fluid-filled cavities of the brain, as represented by Gregor Reisch (*c*.1467–1525). Reisch suggested that images pass through a hierarchical chain of ventricles: the eyes input information to the first set, the *sensus commensis*, and this information is then relayed to the second and third sets of ventricles for progressively higher thought processing.

WELLCOME LIBRARY, LONDON

Figure of the brain with both membranes stripped off (right)
De humani corporis fabrica libri septem
Andreas Vesalius, Basel, 1555
Wood engraving in book

Up until the 19th century, the functions of the cerebral cortex were little understood and scant attention was therefore given to the accurate depiction of its 'convolutions'. Despite this, the engravings commissioned by pre-eminent 16th-century anatomist Andreas Vesalius (1514–1564) portray in detail this mysterious bundle of folded 'grey matter'. Vesalius theorised that the mental faculties lay in the substance rather than the ventricles of the brain but admitted ignorance about how they worked. In this dissection image, the protective membranes (meninges) have been retracted and the two hemispheres of the brain are divided.

WELLCOME LIBRARY, LONDON

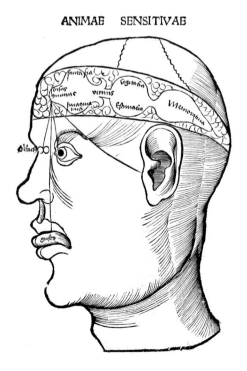

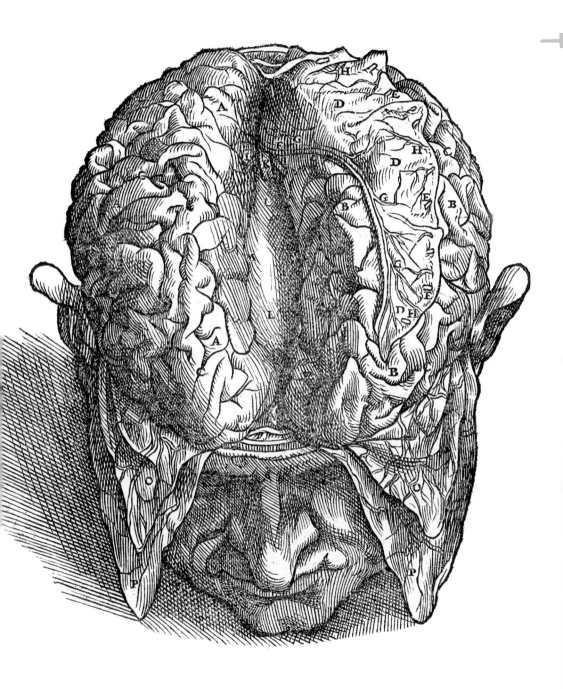

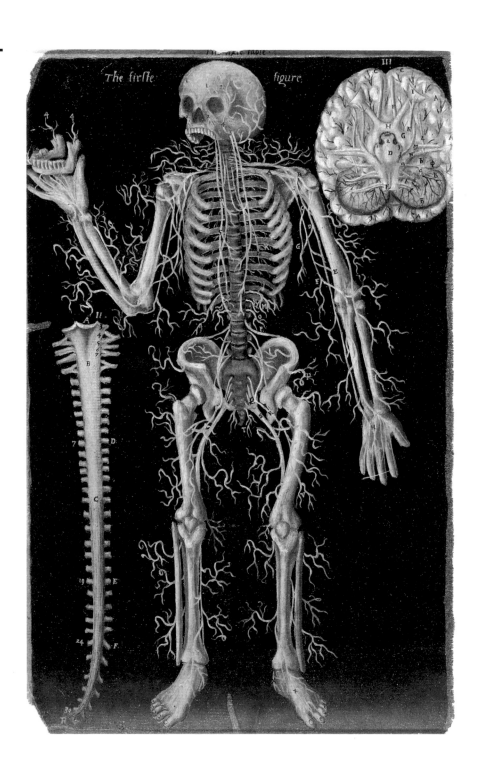

Skeleton, brain, nerves
From John Banister's Anatomical Tables
(tables 5 and 6), *c.*1580
Oil paintings on paper

John Banister (1533–1610) was credited with bringing English anatomy into the light of the Renaissance. Originally part of a bound volume commissioned by Banister, these paintings vividly portray the integral structure of the brain and spinal cord, and therefore the brain in relation to the rest of the nervous system.

SPECIAL COLLECTIONS, UNIVERSITY OF GLASGOW LIBRARY, MS HUNTER 364 (V. 1.1)

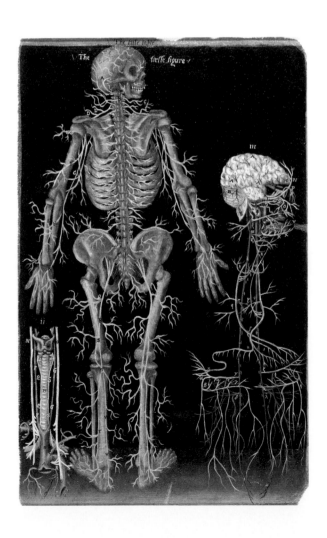

Diagram of the ventricles of the brain and pineal gland (below)
L'Homme de René Descartes et un traitté de la formation du fœtus
Paris, 1664
Wood engraving in book

Following Galen, the Greco-Roman surgeon and philosopher (129–199/217 CE), 17th-century attention concentrated on the ventricular system of the brain. The theorist René Descartes (1596–1650) maintained that the pineal gland was the seat of the soul and the governing centre of the human body, regulating the 'spirits' that connected via the ventricles to the muscles of the body. In this diagram of a transverse section of the brain, the pineal gland is indicated by the letter H and the areas B and C show the nerves running towards the ventricles (A).
WELLCOME LIBRARY, LONDON

Internal parts of the head, including the meninges, the cerebrum, the cerebellum, and the medulla oblongata (right)
Bibliotheca anatomica, medica, chirurgica &c.
Daniel Le Clerc and Jean-Jaques Manget, 1711–14
Copperplate engraving in book

This volume is composed of various descriptions of different body parts, each presented by one or more eminent physicians or chirurgeons (surgeons), and was issued in monthly parts. The engraving of scientific and technical imagery became a highly specialised craft with the boom in 'Enlightenment' publishing during the 18th century. Tables VII and VIII show a series of dissected brain parts laid out in sequence, demonstrating a high level of anatomical detail and a recognition of the relationships between each element.
WELLCOME LIBRARY, LONDON

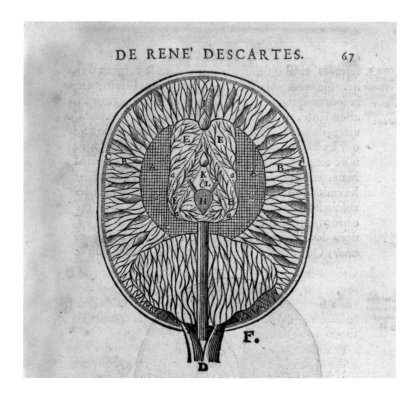

MAPPING/MODELLING 67

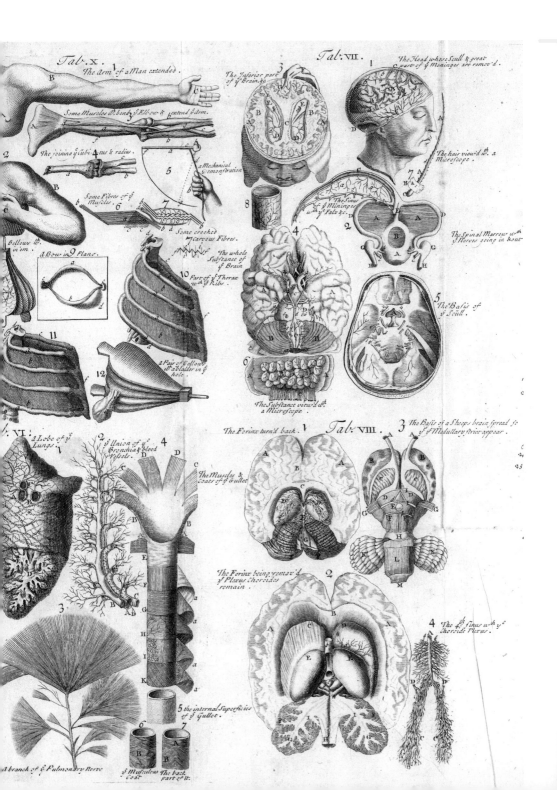

The hemispheres separated to exhibit the vertical walls, falx and corpus collosum (below)
Anatomy of the Heart, Cranium and Brain
Alexander Ramsay, 1813
Hand-coloured print in book; cutaway paper

The anatomist Alexander Ramsay (1754–1824) was proficient at dissection and described himself as having "only two superiors as an anatomist – God Almighty and John Hunter" (considered to be the founder of 'scientific surgery'). He made his own illustrations and, for this volume, engraved his own plates. The pages themselves appear to have been adeptly dissected to reveal the intricate layers of skin and bone, tissue and blood vessels – perhaps also a reference to Ramsay's own anatomical skills.

WELLCOME LIBRARY, LONDON

The Anatomy of the Brain (right)
Charles Bell, 1823
Watercolour in book

Distinguished British anatomist and surgeon Sir Charles Bell (1774–1842) published the first edition of *The Anatomy of the Brain* in 1802. Bell undertook significant work on the localisation of brain function in the cerebrum (the largest portion of the brain, consisting of folded bulges called gyri) and the cerebellum. This plate shows the "general anatomy and subdivisions of the brain". Particular attention has been given to the accurate portrayal of the gyri and the mode of posing the head for dissection.

WELLCOME LIBRARY, LONDON

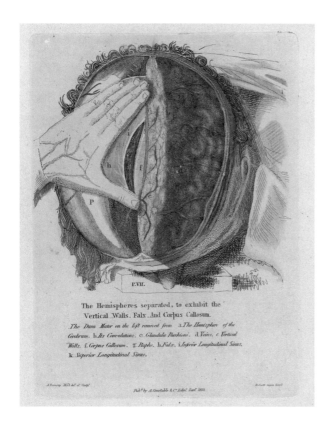

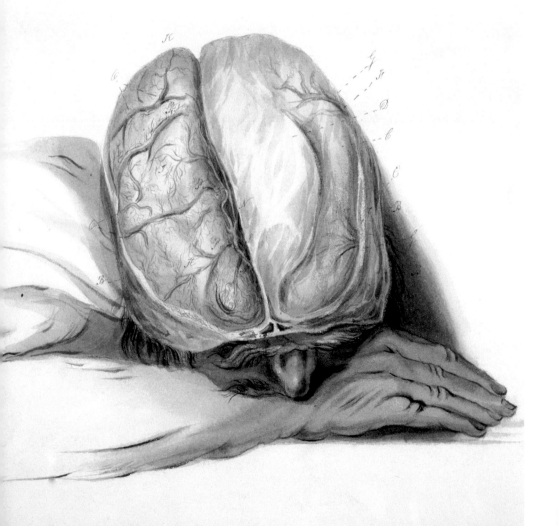

Alligator brain (left)
and dog brain (right)
Ziegler studio, 1887
Wax models with wooden handles

Researchers have long studied animal brains, endeavouring to identify unique features of the human brain that could account for its superior mental abilities. All mammals have the same basic brain structure: the main physical difference is the size and proportion of the different components of the brain. Anatomically, the human brain is very similar to that of other primates, whereas reptiles have a very small cerebrum and rely mostly on their brain stem, which regulates heartbeat and breathing.

THE HUNTERIAN, UNIVERSITY OF GLASGOW

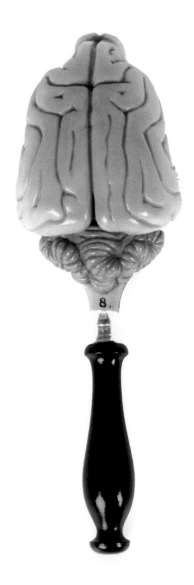

Dissected recumbent human head
Joseph Towne, c.1850
Wax

Joseph Towne (1806–1879) was one of the most skilled and prolific makers of wax moulages (medical models) of the 19th century, and also a talented sculptor in marble. He worked exclusively for Guy's Hospital in London. He embraced new technologies such as stereoscopic photography to help him achieve realism; however, like the illustration to Charles Bell's *Anatomy of the Brain*, this model conforms to the rhetoric of classical posture, which had prevailed in anatomical art from the time of Vesalius.
GORDON MUSEUM, KING'S COLLEGE LONDON

MAPPING/MODELLING

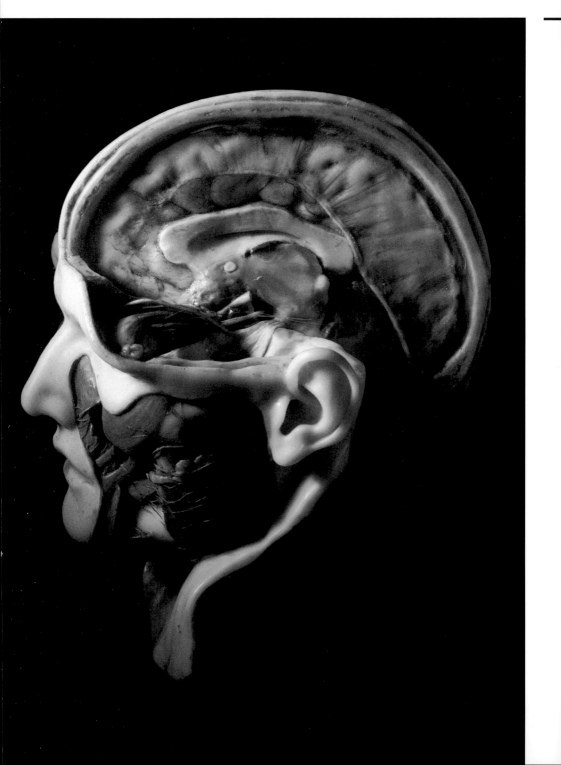

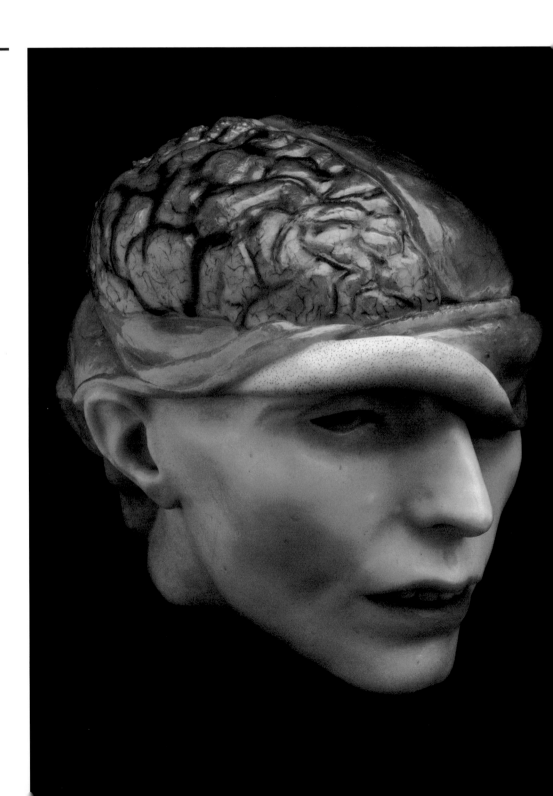

Anatomical model of human head (left)
Possibly German
19th century
Wax

Adopting a pose also favoured by Joseph Towne, the cranium is shown removed and the scalp folded over to reveal the underlying surface of the cerebral cortex with the dura mater (the protective membrane, part of the menginges) still covering the left hemisphere. The model may have been made to teach neuroanatomy to medical students or, given its unsettling facial expression, for medical exhibitions open to the general public.
SCIENCE MUSEUM, LONDON

Cast of brain section to demonstrate distribution of nerve elements (below)
William Macewen, 1893
Plaster

Scottish surgeon William Macewen (1848–1924) was a pioneer in modern brain surgery and one of the first to operate successfully on a brain tumour after having precisely determined its location by assessing the patient's epileptic symptoms. Macewen amassed a large collection of clinical photographs, specimens and plaster casts, many of which he produced himself and used for teaching purposes.
THE HUNTERIAN, UNIVERSITY OF GLASGOW

Corrosion cast of blood vessels in the brain
1980s
Resin model

The entire vascular (blood vessel) system of a whole animal or organ, down to the smallest capillaries, can be injected with liquid plastic which fills the blood space and rapidly solidifies. The surrounding tissue is then corroded away with an acid or alkaline solution, leaving a hardened cast that accurately represents the form of the original vascular system. With the finest casts made from modern low-viscosity resin, the results can be inspected with electron microscopes, and their geometry mathematically analysed.
GORDON MUSEUM, KING'S COLLEGE LONDON

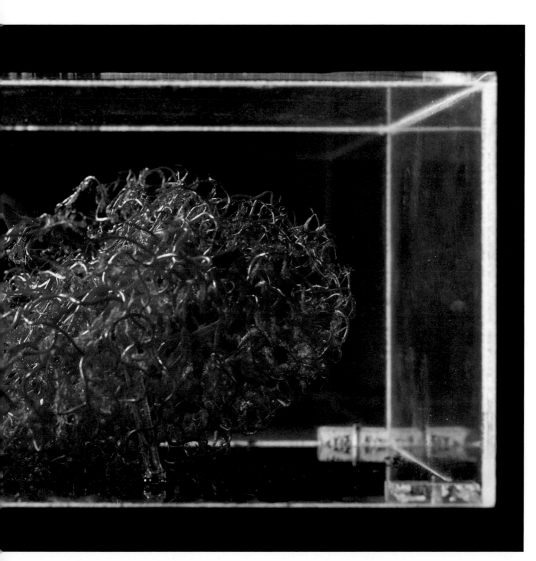

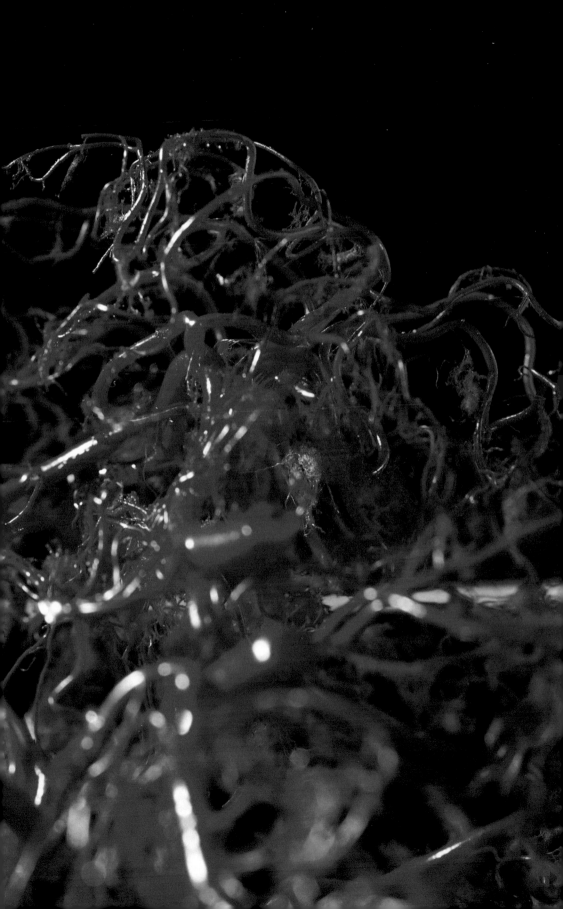

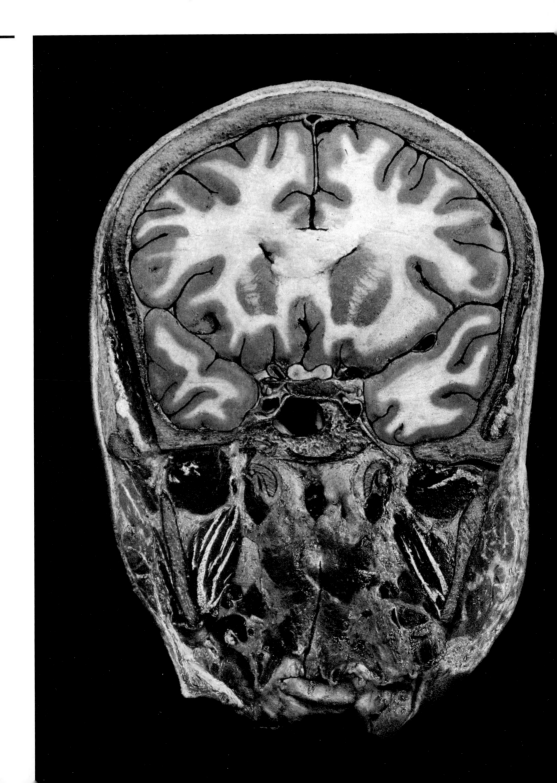

Atlas of Head Sections
William Macewen, 1893
Photogravures in book

Macewen's thorough examination of neuroanatomy culminated in this extraordinary atlas containing 53 copperplate engravings from his own photographs of frozen head sections. The results are reminiscent of modern-day CT (computed tomography) imaging of the brain. In the pre-X-ray age, the atlas was intended to be used by surgeons as a reference tool prior to, and during, surgery. Macewen's knowledge of cranial anatomy was important in developing his successful treatment of brain abscesses.
WELLCOME LIBRARY, LONDON

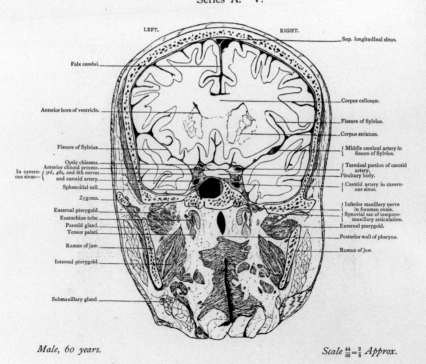

Atlas: There and Here
Andrew Carnie, 2012
Video installation

Andrew Carnie's work explores themes of change and mortality and the haunting imagery produced by the biomedical sciences. 'Atlas: There and Here' is based on William Macewen's *Atlas of Head Sections* (1893) and reflects Carnie's interest in the act of anatomical slicing. It interleaves images derived from the book with recent photographs of Macewen's original specimens at the Hunterian in Glasgow. The video proceeds through strange-looking anatomical renditions of the head, occasionally returning to the more familiar domains of the medical textbook and the museum – what Carnie calls "the sphere of the living".

COURTESY OF THE ARTIST AND GV ART, LONDON.

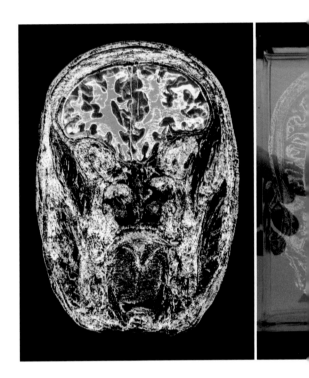

MAPPING/MODELLING 83

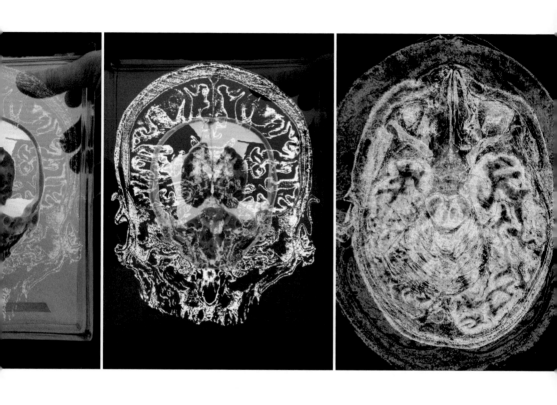

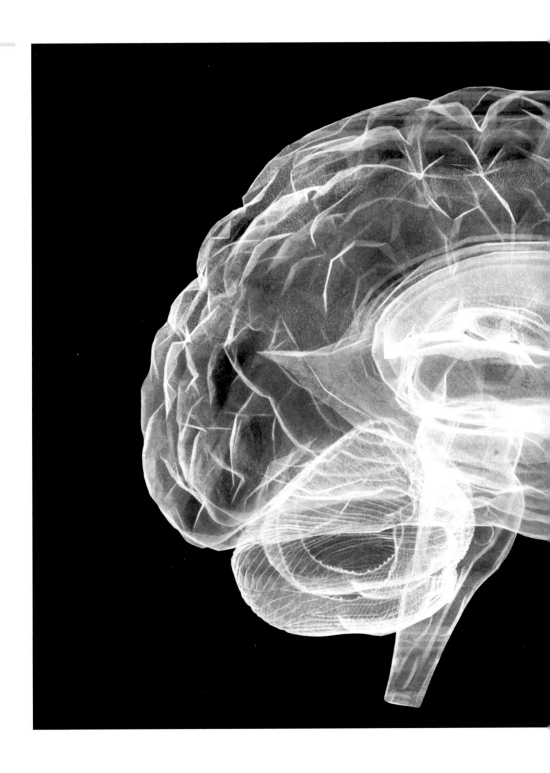

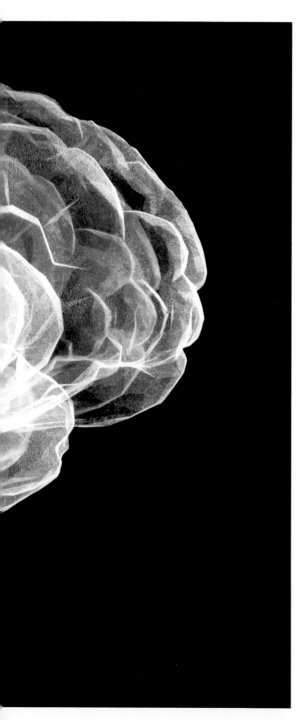

My Soul
Katharine Dowson, 2005
Laser-etched lead crystal glass

Using her own MRI (magnetic resonance imaging) scans, made during research into dyslexia, Katharine Dowson laser-etched a 3D representation of her brain onto the inside of a glass block. The choice of material plays with perception, while the etched brain structure suggests a real, physical object. Its transparency provides an unobstructed view into the interior of the brain as well as the shape of the organ that enables self-consciousness.
COURTESY OF THE ARTIST AND GV ART, LONDON

Self-portrait of Santiago Ramón y Cajal
1895

Santiago Ramón y Cajal (1852–1934) is widely regarded as the father of modern neuroscience. As a child, Cajal showed a precocious talent for drawing and painting, but his father disapproved of an artistic career, so he followed the family tradition and became a doctor. He jointly won the Nobel Prize in 1906, with his rival Camillo Golgi (1843–1926), for their contribution to the understanding of the structure of the nervous system. Cajal also found time to write science fiction and to father a successful family.

Golgi had discovered that approximately one in every thousand nerve cells absorbed a dye, revealing its structure. This 'Golgi stain' was a breakthrough in the visualisation of brain tissue, which previously had remained pale and enigmatic under the microscope. Both Golgi and Cajal made exquisite drawings of what they witnessed using this technique, but it was Cajal's theory, that the brain comprises individual nerve cells rather than a continuous mesh of filaments, which triumphed. His 'neuron doctrine' still stands.

The story goes that Cajal drew from memory; he certainly used a *camera lucida* and made thousands of drawings on any scrap of paper at hand. His visual intuition enabled him accurately to deduce the flow of impulses through the complex circuitry he saw. There is a distinct Art Nouveau spirit in the sinuous, graphic enchantment of his 'garden of neurology'.

COURTESY OF CAJAL LEGACY, INSTITUTO CAJAL (CSIC), MADRID

Ganglionic cells of the trigeminal nucleus of the mouse (below)
Microscope slide by Santiago Ramón y Cajal, c.1900; photomicrograph by Virginia García Marín
COURTESY OF CAJAL LEGACY, INSTITUTO CAJAL (CSIC), MADRID

Parasagittal section of the cerebellum (detail) (right)
Santiago Ramón y Cajal, 1894
Ink drawing on paper
COURTESY OF CAJAL LEGACY, INSTITUTO CAJAL (CSIC), MADRID

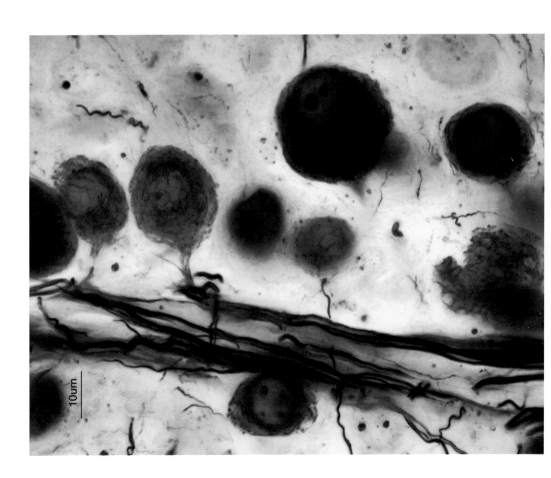

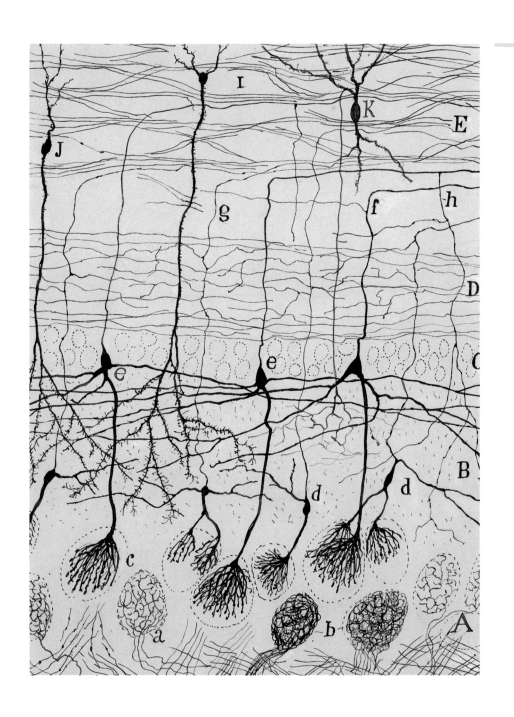

Brainbow Mouse
Jean Livet, Joshua R Sanes and
Jeff Lichtman, 2007
Digital photomicrograph

Modern biologists are able to make living cells fluoresce by altering them with genes taken from coral and jellyfish. However, even after marking them in this fashion, visually disentangling individual neurons from their neighbours remains difficult. In 2007 a team at Harvard University made a breakthrough by developing the 'Brainbow' mouse. Their technique enabled whole arrays of similar neurons to be discriminated from each other because each one expresses different combinations of a palette of coloured proteins. Photographs are then digitally re-coloured to enhance the contrast between cells and reveal spectacular maps of the complex interconnections within the nervous system.

LICHTMAN LAB, HARVARD UNIVERSITY

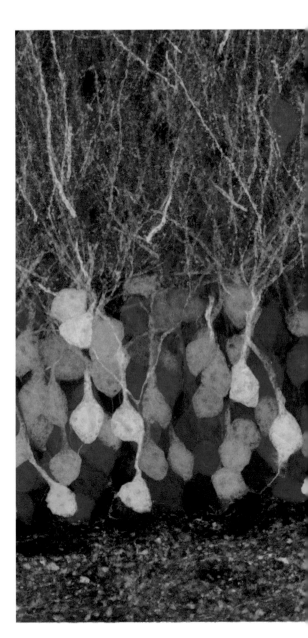

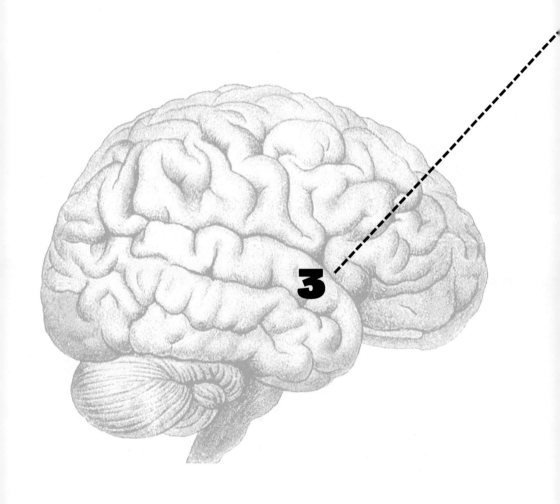

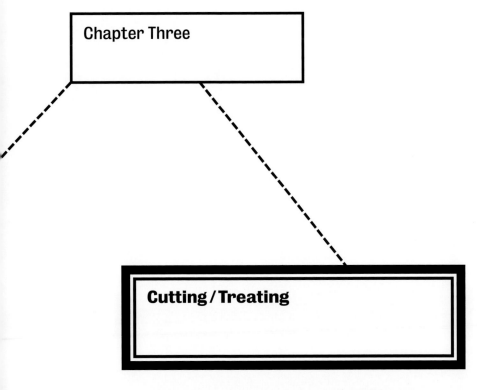

Chapter Three

Cutting / Treating

Undaunted by the delicate nature of the brain, humans have been trying since prehistoric times to intervene beneath the skull – to relieve the pressure of bleeding, to excise tumours, to treat more general ailments, or even to achieve a drug-like high. Some of the instruments and techniques for accessing the brain through its protective layers of skin, bone and membranes remain fundamentally similar to this day. Although the brain itself often shows remarkable resilience to surgery, brain tumours are still very dangerous and much more research on them is needed.

In the past century, neurosurgeons have used X-rays, scanning and 3D geometrical devices to help them locate the source of the problem while minimising collateral damage. Accidents and warfare through the ages have also provided many cases of wounded and traumatised brains from which clinicians and scientists have learned. The vital importance of electricity to nerve function has been known since the 18th century, and it has been used to treat depression and other neurological disorders – often effectively, though with varying degrees of theoretical understanding, precision and humanity.

**Me just before brain surgery
London hospital 1996**
Corinne Day, 2000
C-type print mounted on aluminium

In 1996, the fashion photographer Corinne Day (1962–2010) was diagnosed with a brain tumour, at which point she turned the camera on herself, having given instructions for her ensuing medical journey to be documented in all its grim reality. The apparatus attached to her head is a stereotactic frame used to help guide the surgeon to a target within the brain, or to perform a biopsy for diagnosis.

WELLCOME COLLECTION, LONDON

CUTTING/TREATING 95

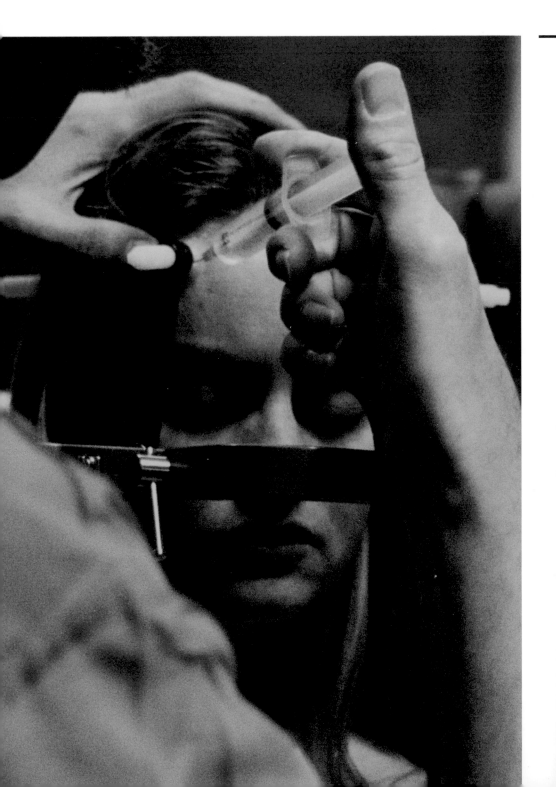

Bronze Age skull from Jericho, West Bank, Palestinian Territories (below)
2200–2000 BCE

This skull shows four separate holes made by the ancient surgical process of trephination. Also known as trepanation, or trepanning, the practice of cutting through to the surface of the brain might be carried out to treat a variety of medical conditions. Some of these, such as pressure within the cranium caused by infection or bleeding on the brain, would genuinely be alleviated by this method. This apparent success may have been explained as the release of evil spirits from the body.
SCIENCE MUSEUM, LONDON

Trephined skull fragment (right)
Dr T Wilson Parry, 1913

Thomas Wilson Parry (1866–1945) was an English doctor who made and experimented with instruments of a type that would have been used to perform trephinations in the Neolithic period. Parry collected skulls from around the world and drilled into them with a wide range of materials including glass, flint, sharks' teeth, seashells and obsidian (volcanic glass). He discovered, perhaps unsurprisingly, that each material and instrument makes a different type and size of hole, and takes different amounts of time to penetrate.
SCIENCE MUSEUM, LONDON

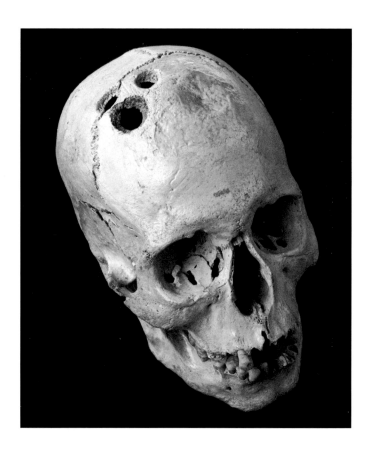

CUTTING/TREATING 97

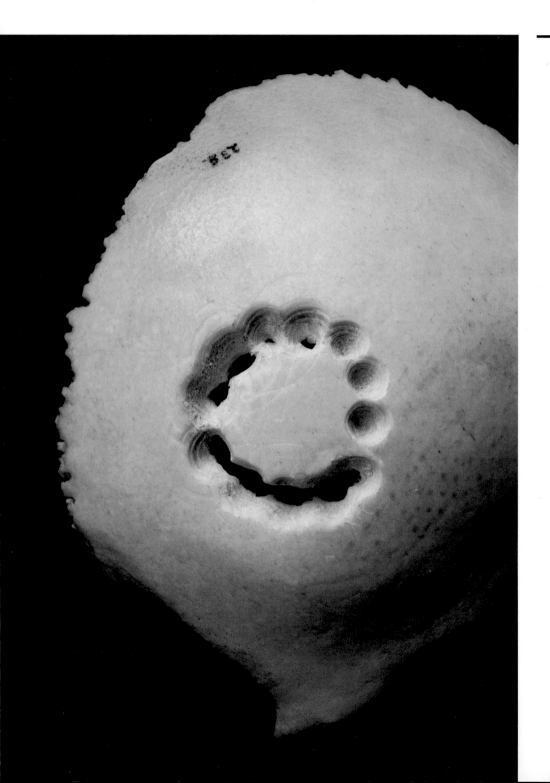

Cigar box containing oyster shells similar to those used in trephination (left)
Dr T Wilson Parry, 1900–1915
Wood, cotton string, metal, shell
SCIENCE MUSEUM, LONDON

Trephination and cranial surgery instruments (various)
Trephines are the surgical devices used for trephination or trepanning. The basic practices and tools have remained largely unchanged for centuries.

Trephine (below, top left)
Wooden handle. shark-tooth blade

Bone-dust brush (below, top right)
19th century
Bristle with wooden handle

Cranial file (below, bottom right)
Before 1870
Steel with ebony handle

Trephine (below, bottom left)
19th century
Brass, steel, wood
SCIENCE MUSEUM, LONDON

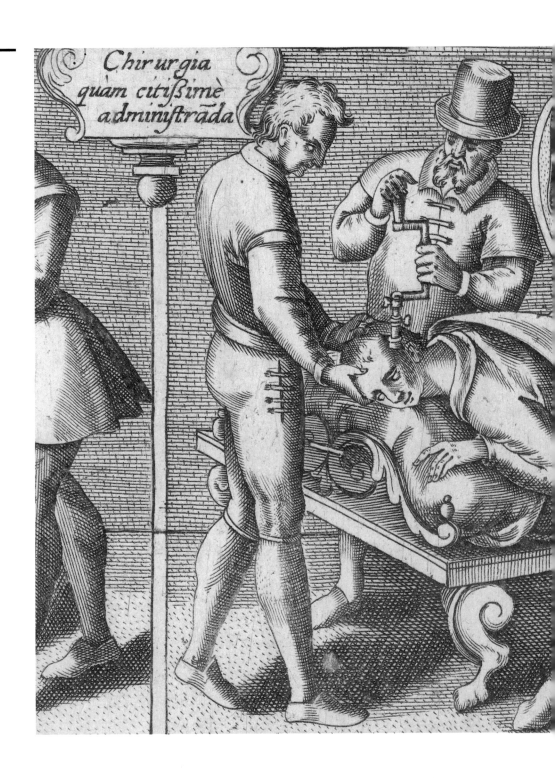

Performing a trephination procedure
Le chirurgie Françoise recueillie des antiens médecins et chirurgiens
Jacques Guillemeau, 1594
Copperplate engraving in book

The French surgeon Jacques Guillemeau (1550–1612) is credited with improving the trephine and invented a successful technique for operating on aneurysms (ballooning arteries). This engraving from his textbook, supposedly based on the methods of the ancients, possibly shows field surgery from one of the Spanish army campaigns in Flanders that Guillemeau attended. One patient is being trephined while another, clearly apprehensive, is having his scalp prepared.
WELLCOME LIBRARY, LONDON

Photograph of Dr Harvey Cushing with Professor Otfrid Foerster and patient (below)
Walter Willard Boyd, 1930

Harvey Cushing (1869–1939) is widely regarded as the founder of modern brain surgery. By combining a careful approach with new research and technology, such as X-rays and the monitoring of blood pressure, he was able to drive surgical mortality rates down to less than 10 per cent, at a time when most other neurosurgeons were losing about 30–45 per cent of patients. He bequeathed over 15 000 clinical portraits of patients and most of the specimens from his career in a 'tumour registry' at Yale University, where he was Professor of Neurosurgery.
CUSHING/WHITNEY MEDICAL LIBRARY, YALE UNIVERSITY

Pre-operative photograph of female patient with craniopharyngioma (right)
1919

Many of the patients in these portraits presented with much more advanced tumours than would normally go unchecked today. The 15-year-old subject of this photograph suffered years of headaches, nausea, convulsions, restricted development and impaired vision before being referred to Dr Cushing. She was in and out of hospital for the next 12 years, though the final letter in her file, from her father in 1931, strikes an optimistic note and thanks Cushing for his care.
CUSHING/WHITNEY MEDICAL LIBRARY, YALE UNIVERSITY

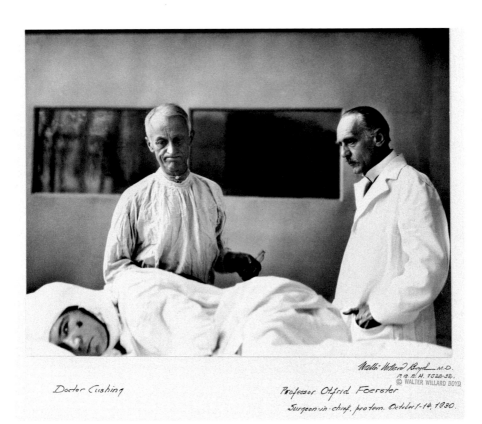

Pre-operative photograph of male patient with pituitary adenoma (below)
1914

An excess of growth hormone caused by a tumour of the pituitary gland in the brain can result in acromegaly and gigantism, where the patient grows very tall and suffers a coarsening of the facial features, enlarged hands and feet, and thickening and wrinkling of the scalp. Pituitary conditions became Cushing's specialism. Unfortunately, this patient died after his second operation; his skeleton was preserved and photographed in comparison with a normal specimen.
CUSHING/WHITNEY MEDICAL LIBRARY, YALE UNIVERSITY

Photograph of patient following operation for parasagittal meningioma (right)
1930

This portrait shows a patient two weeks after an operation for a tumour of the membrane between the skull and the brain. He is reported to be "perfectly well in every way". In 1910, Cushing had performed the first ever successful meningioma operation, on General Leonard Wood, Chief of Staff of the US Army. Wood went on to stand for the Republican nomination for the US presidency, but the tumour recurred and in 1927 he died following another operation by Cushing.
CUSHING/WHITNEY MEDICAL LIBRARY, YALE UNIVERSITY

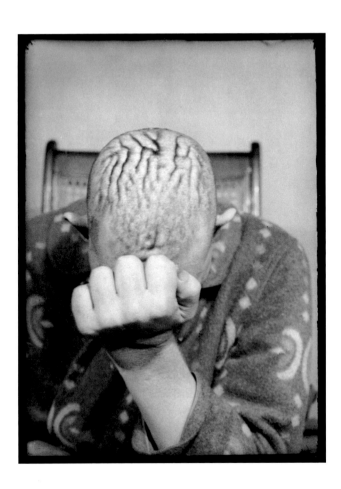

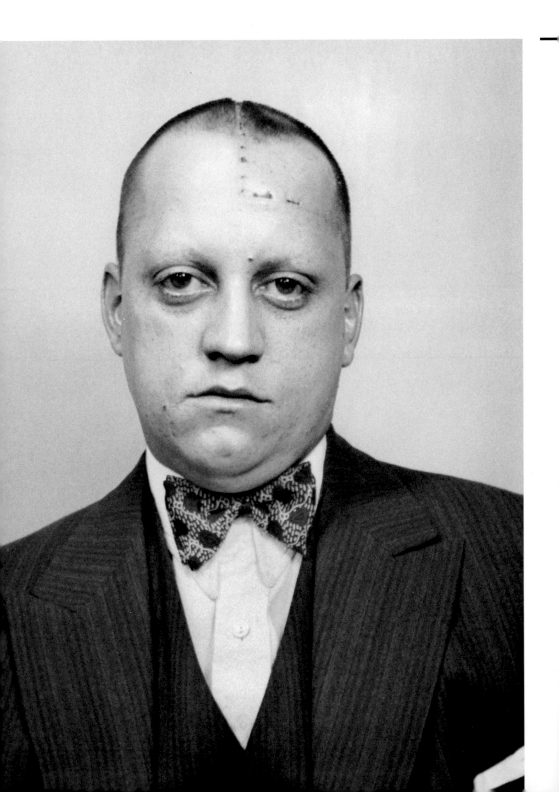

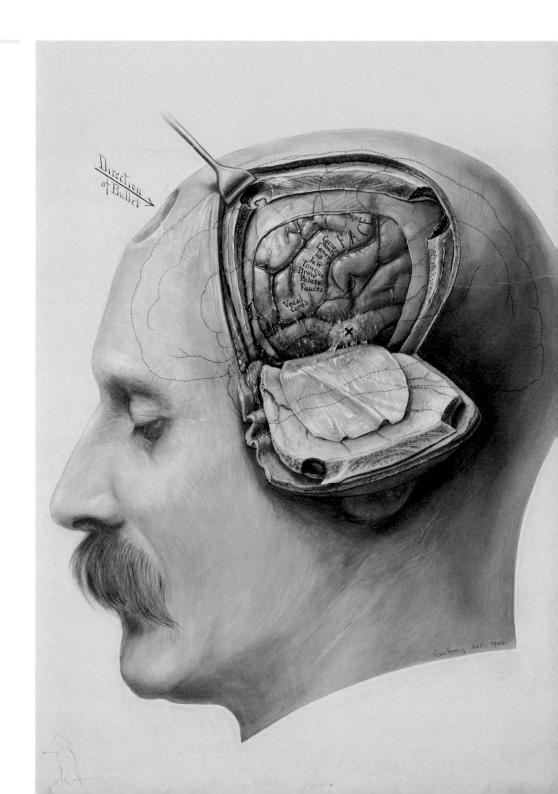

Anatomical drawing showing operation for a gunshot wound
Harvey Cushing, 1906

Cushing was a gifted artist, as well as a writer (in 1926 he won the Pulitzer Prize for his biography of his mentor, the Canadian physician Sir William Osler). This illustration was for a surgical monograph, reflecting recent research into the mapping of the motor cortex. Cushing's technique shows the influence of his friend and colleague, the German medical illustrator Max Brödel. The profile is Osler's, and the subject presages Cushing's later service as a surgeon in France in World War I, where he tried in vain to save the life of Osler's wounded son.
CUSHING/WHITNEY MEDICAL LIBRARY, YALE UNIVERSITY

Cerebrospinal rhinorrhoea
Audrey Arnott, 1950
Pencil, grey wash and white goache

Audrey Arnott (1901–1974) graduated from the Royal College of Art before working as a medical illustrator for the neurosurgeon Sir Hugh Cairns. Cairns arranged for her to train with Max Brödel at the prestigious Johns Hopkins Hospital in Baltimore, where she learned the technique of drawing with Ross-board (a stippled paper) and carbon dust applied with dry brushes. Arnott returned to work with Cairns at Oxford, where she made many sketches recording his operations. Even her formal illustrations often depict the faces of patients with considerable sympathy.
WELLCOME LIBRARY, LONDON

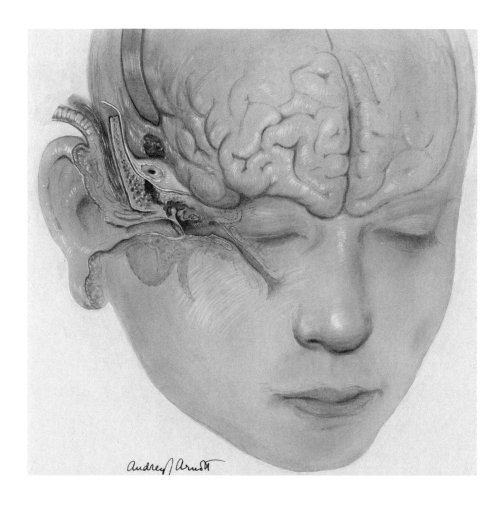

CUTTING/TREATING 109

Hemispherectomy in an adult patient: four figures
Audrey Arnott, 1960
Carbon dust drawing on paper

A drastic operation to cut the corpus callosum joining the two halves of the brain, sometimes used in cases of severe epilepsy.
WELLCOME LIBRARY, LONDON

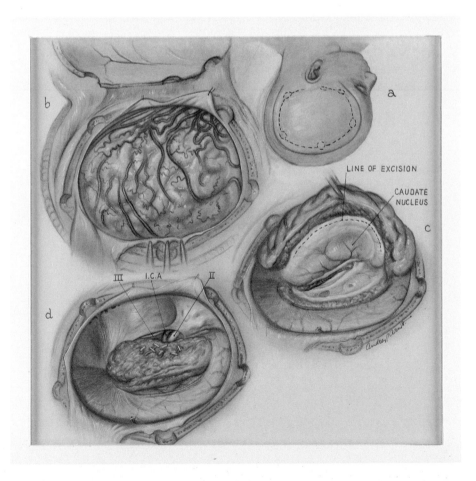

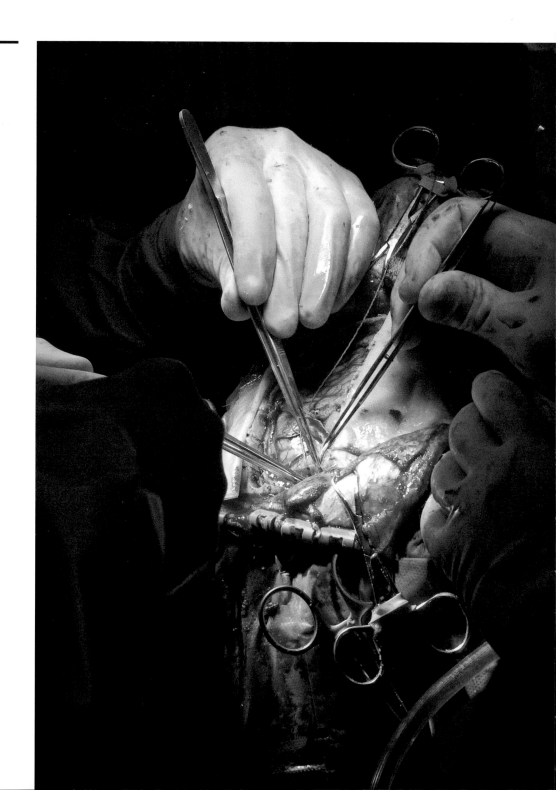

Professor Dennis Spencer and team performing brain surgery
Terry Dagradi, 2010

Dennis Spencer is Professor of Neurosurgery at Yale University, and custodian of the Cushing legacy. These photographs of him leading a surgical team were taken by the curator at the Cushing Center, Terry Dagradi. What most distinguishes modern neurosurgery from that of Cushing's day is the availability not only of antibiotics to control infection, but also of instantaneous scanning imagery to help guide the surgeon and minimise the intervention required.
CUSHING CENTER, YALE UNIVERSITY

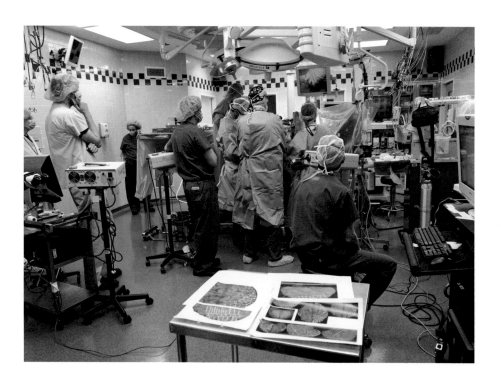

'Jedi' helmet, used with cryogenic MRI machine
1980
Metal (copper) and plastic

These helmets were designed as an experimental device in order to obtain the best possible scans of a child's brain. The coils are 'aerials' for picking up MRI (magnetic resonance imaging) signals, which generate real-time pictures from the magnetic behaviour of water molecules in the human body. MRI does not expose the body to radiation and can create images of the soft tissue more effectively than X-ray-based methods.

SCIENCE MUSEUM, LONDON

CUTTING/TREATING 113

Memory of a Brain Malformation
Katharine Dowson, 2006
Laser-etched lead crystal glass

Katharine Dowson's work is based on her cousin's cerebral angiogram – an X-ray test which highlights the blood vessels in the head. Arteriovenous malformation (AVM) is caused by too direct a connection between arteries and veins, producing a fragile tangle of blood vessels called a nidus, from the Latin for 'nest'. This, along with the swollen blood vessels supplying it, can clearly be seen in the etched image within the glass. In this case the AVM was successfully treated by laser – also the means for creating this work.
COURTESY OF THE ARTIST AND GV ART, LONDON

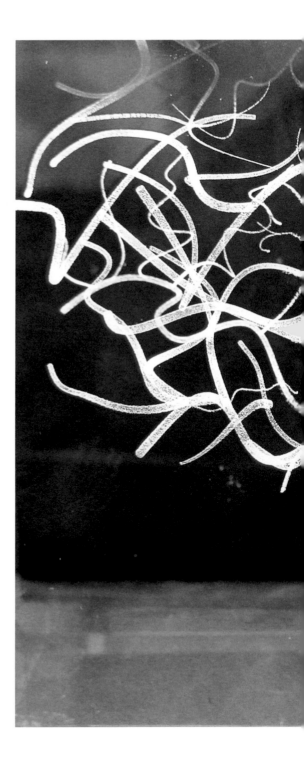

CUTTING/TREATING

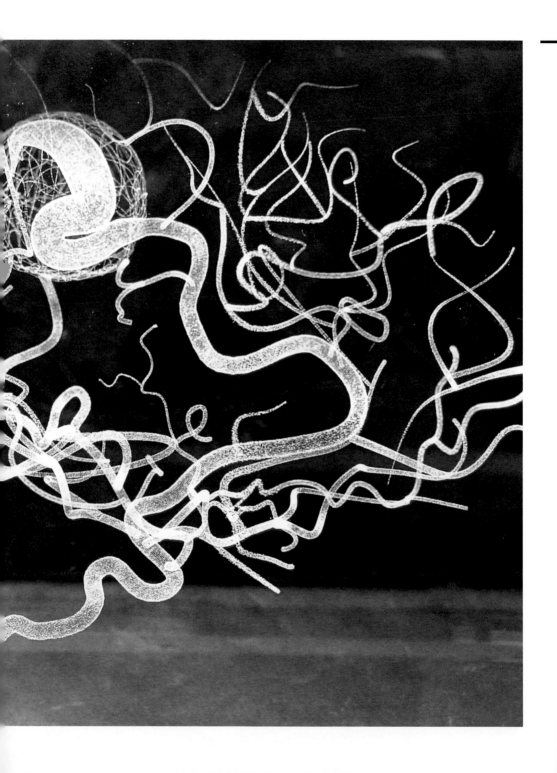

Hearing Voices
Susan Aldworth, 2000
Watercolour on paper

This series of drawings comes from Susan Aldworth's work as artist-in-residence at the Royal London Hospital in Whitechapel. They are based on scans taken during treatment of a four-year-old boy for a fistula (an artery feeding directly into a vein) at the base of his brain. He had been complaining of hearing Martians talking in his head. Aldworth was observing throughout and recorded in her diary that the boy was terrified of the anaesthetic before a catheter was inserted into his brain: "'I don't want the gas in my face' he screams over and over again". Once the procedure was underway, she found it a relief from the tension and anguish to be able to concentrate on her drawing.

COURTESY OF THE ARTIST AND GV ART, LONDON

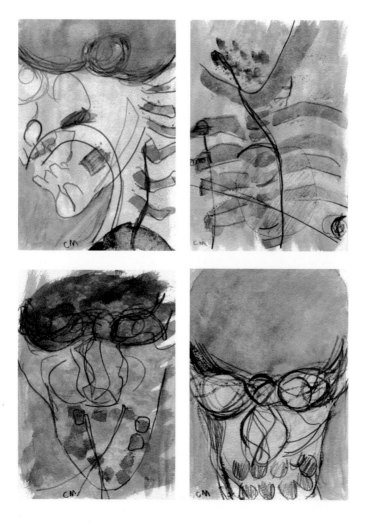

Electroconvulsive Therapy Machine, model R1135 (below)
1945–60
Wood, metal, plastic, vulcanite

Developed in Italy in 1938, electroconvulsive therapy (ECT) is a controversial psychiatric treatment, primarily used for cases of severe depression. An ECT machine transmits electric currents to the brain in timed pulses causing a convulsion or seizure, thought to 'shock' patients out of their mental disorder. This example comes from St Audry's Hospital in Suffolk, instituted in 1829 as the Suffolk County Asylum for Pauper Lunatics and closed in 1993.
SCIENCE MUSEUM, LONDON

Electroencephalogram (right)
Burden Neurological Institute, Bristol, 1958
Ink trace on paper

Although rooted in 19th-century neuropsychiatry and experimental psychology, human electroencephalography (EEG) was officially pioneered by Hans Berger (1873–1941) in the 1920s, when he published the first record of human 'brain waves' using electrodes placed on the surface of the head. This recording demonstrates the results of an electrical stimulation experiment conducted on an epileptic patient by the Burden Neurological Institute, which specialised in the human nervous system and neurological disorders.
SCIENCE MUSEUM, LONDON

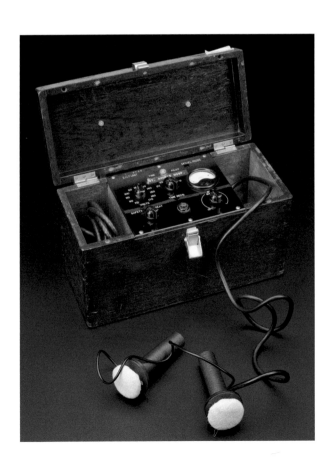

CUTTING/TREATING

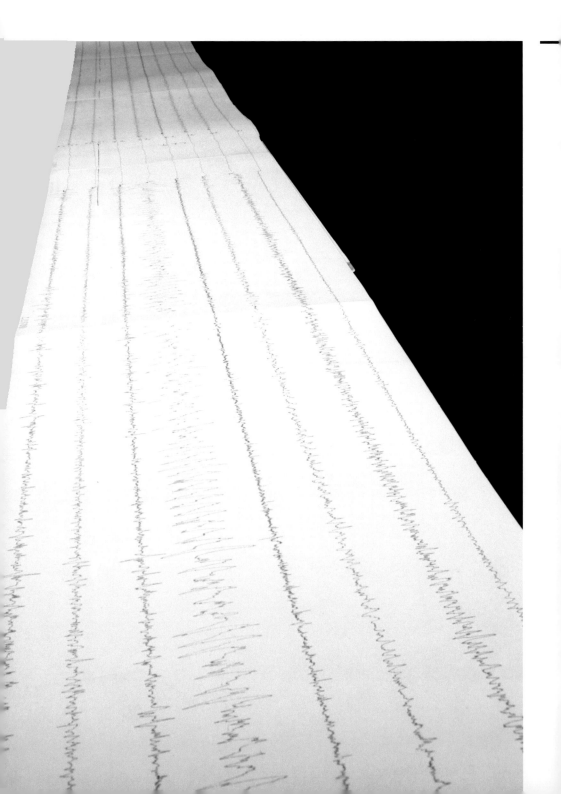

CUTTING/TREATING 121

Me, Myself, I
Allan Beveridge, 1987
Drawing on scraperboard

This work is part of the collection of art by patients at the Royal Bethlem Hospital in London. Dating from the Middle Ages, Bethlem is the probably the oldest mental hospital in the world, and has itself featured heavily in art and public consciousness, mostly because of its fearsome reputation. Today it is at the forefront of humane psychiatric treatment.

Writing in 1991, Allan Beveridge said of his work:

There is a symbolic structure to the picture. The clowns' faces represent aspects of myself. The chess board and chess men represent intelligence and the exposed brain with the dagger represents the ill mind. The rest of the picture is biographical: the arrangement of needles, pills and cigarettes used to promote harmony, and at the bottom left of the picture, the uneven surface where the pills are being straightened out by the injection (Modecate, which kept me well for years).

BETHLEM ROYAL HOSPITAL ARCHIVES AND MUSEUM SERVICE

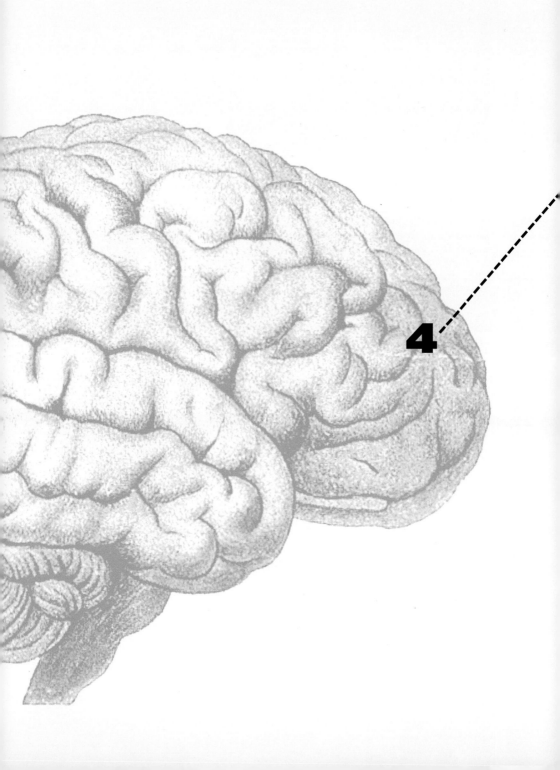

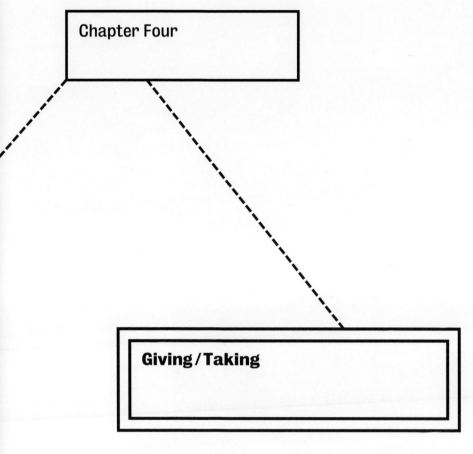

Chapter Four

Giving/Taking

Since appropriate preservation techniques were developed in the late 18th century, the collecting of specimens has been as important to brain science as it has to other branches of natural history, anatomy and pathology. Biologists sought the brains of humans and other species for comparative study and to represent different classes or categories as well as anomalies and diseases. However, they often failed to seek the prior consent of donors or their families.

In Nazi Germany, neurologists were involved in the mass murder of supposed inferiors; some made illustrious careers from studying the preserved brains of their victims. Many countries in Europe and North America have revised the codes and laws governing the preservation and supply of human tissue in recent years. In the context of ageing populations and an epidemic of neurodegenerative disorders, scientists are now encouraging members of the public to pledge their remains to brain banks whether they are healthy or not.

Letters on the brain of baby Donald McGee
Dr Anita Newcomb McGee, 1896
Photographic portrait, letter and pencil sketch on paper

Donald McGee was the second child of Anita Newcomb McGee (1864–1940), an American doctor who later became the only female Acting Assistant Surgeon in the US Army. These items come from the file of her correspondence with the renowned brain anatomist Burt Green Wilder at Cornell University in upstate New York, to whose extensive collection she donated the brain of her son after he died of meningitis at the age of nine months. Her letters show, through overtones of professionalism, a desire to turn her unfathomable loss into scientific and medical progress by bequeathing his brain for expert examination: "I want him to benefit the world in some way."

BURT GREEN WILDER PAPERS, DIVISION OF RARE AND MANUSCRIPT COLLECTIONS, CORNELL UNIVERSITY LIBRARY

Ans'm'ed
June

South Atlantic City, N.J.
24 June 1896

Dear Prof. Wilder,

My baby boy is in a critical condition from meningitis supervening on bowel trouble. He is nine months old. I came here from Washington in hopes of benefiting him. Would his brain be of any value to you? If so, please send at once explicit directions. I want him to benefit the world in some way if possible.

Sincerely yours
Anita Newcomb McGee

Donald McGee
Profile of top of head — from memory. No "bumps". Chin very retreating at birth, growing less so with age.

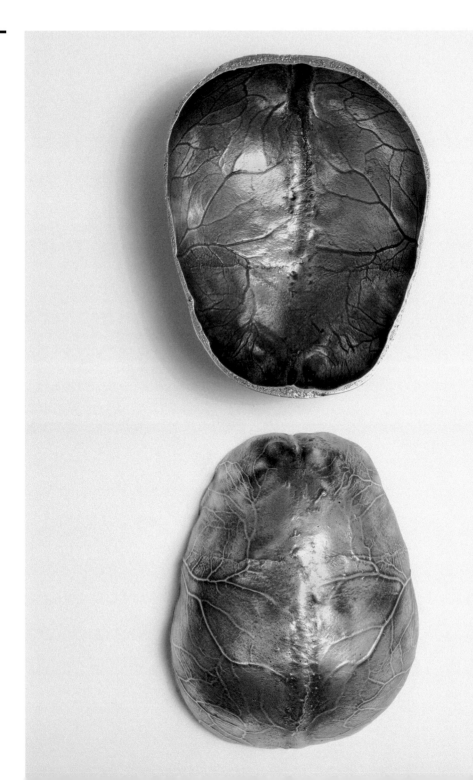

From Within
Annie Cattrell, 2006
Silvered bronze cast

Annie Cattrell has a fascination with metamorphosis and the hidden structures, systems and landscapes of the human body. These casts transform the protective shell of the cranium into a pair of precious objects, revealing the impression of the brain on the interior surface of the skull – an anatomically sensitive interface. They were taken from a skull held at Guy's Hospital in London, where the cap had been cut off to facilitate removal of the brain.

COURTESY OF THE ARTIST

Post-mortem set (below)
John Weiss, 1801–1900
Wood and brass case, stainless steel
instruments with ebony handles

This set contains a large steel chisel-hammer for prising open the skull cap, a brain knife, a spine chisel, saws, forceps and scalpels. Post-mortem examinations, or autopsies, have been carried out for millennia and usually rely on a fairly basic and traditional set of instruments. As well as determining the cause of death, a post-mortem also identifies any prior disease or injury – this in turn has revealed much about internal symptoms and signs of disease.

SCIENCE MUSEUM, LONDON

Post-mortem pathology: a manual of post-mortem examinations (right)
Henry W Cattell, 1903
Photographic book illustrations

The brain is notoriously difficult to access, and removing it after death requires a series of sharp instruments to penetrate the various layers of skin and bone. This delicate process marks the transition between the brain as 'self' and the brain as an object of study, when it usually becomes physically disconnected from the spine and is completely exposed.

WELLCOME LIBRARY, LONDON

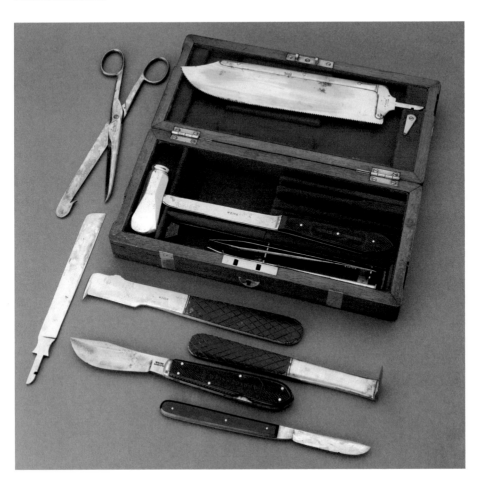

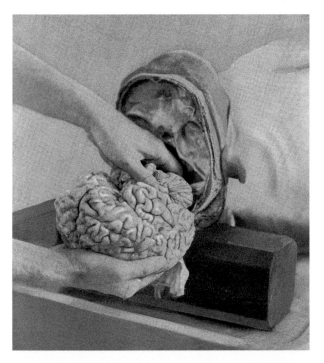

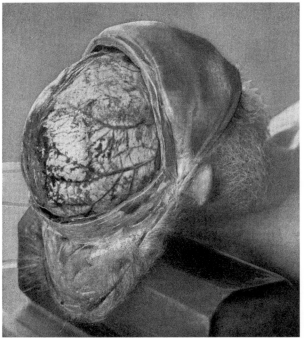

Nervous Tissue Note Panel
David Marron, 2010
Mixed media

The artist David Marron, who also works as a paramedic, assembled this collage in response to a brain dissection he attended at a London tissue bank for GV Art's *Brainstorm* exhibition. He says it is "like an opened note book, revealing thought processes ... The charcoal drawings were made bearing in mind the wealth of anatomical drawings and engravings from the past." Its various artefacts, including "the long knife that passes easily through the brain", are stranded on clinical, white backgrounds like the dissected body parts.

COURTESY OF THE ARTIST, GV ART AND MULTIPLE SCLEROSIS SOCIETY AND PARKINSON'S UK TISSUE BANK AT IMPERIAL COLLEGE LONDON

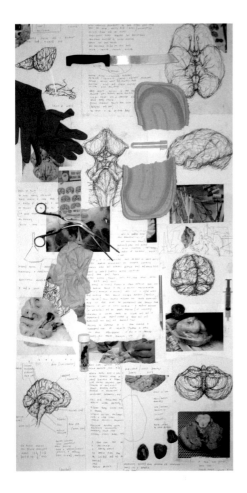

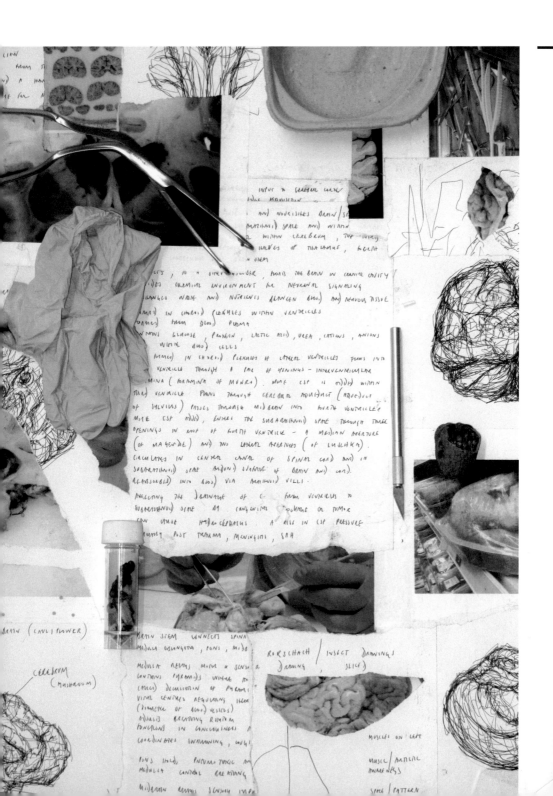

Brain slices
*c.*1950–2010
Dry specimens (human and animal tissue)

The study of thin sections of tissue, or histology, became central to the disciplines of anatomy and pathology with the development of powerful microscopes during the 19th century – indeed, during that period histology became a scientific discipline in itself. These glass slides show some of the range of staining techniques used to highlight different aspects of cell structure, sometimes with contrasting effects apparent to the naked eye. Principally made for teaching purposes from human and other primate material, some of them show signs of wear and tear.

DEPARTMENT OF ANATOMY AND HUMAN SCIENCES, KING'S COLLEGE LONDON

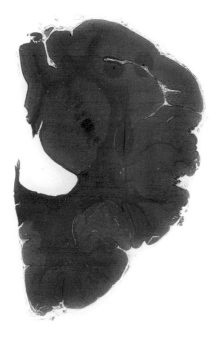
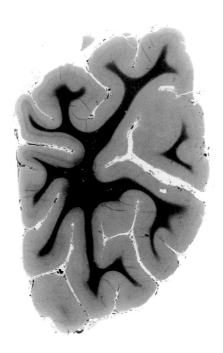

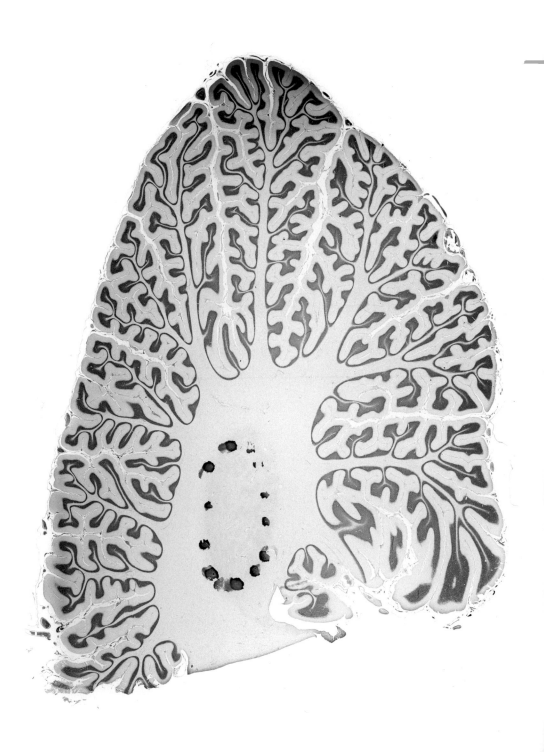

Brain Bank
Martha Henson, 2012
Digital video

Every Wednesday morning at the Imperial College Tissue Bank at Hammersmith Hospital, a few human brains are dissected according to an international protocol designed to facilitate a thoroughgoing pathological report.

The brains have first been 'fixed' by being soaked for a few weeks in formalin-based preservative, and are accompanied by a consent form, usually completed by the donor. Since the Tissue Bank is supported by the Multiple Sclerosis Society and Parkinson's UK, many donations are from people with these debilitating and often tragic illnesses, but control samples of healthy brains are required too.

A basic case history for each numbered, anonymous specimen is read out by a technician, before the pathologists proceed and report evidence of the lesions and abnormalities they can see with the naked eye; many of these, such as lacunae or cavities resulting from stroke, are in addition to the diagnosed conditions. Each major region of the brain is inspected, and then samples of tissue are fixed in wax or deep-frozen for further research. This documentary, filmed for the *Brains* exhibition by Martha Henson of the Wellcome Trust, conveys some of the craft discipline exercised by scientists in their research into neurodegenerative disease.

WELLCOME COLLECTION, LONDON
COURTESY OF MULTIPLE SCLEROSIS SOCIETY AND
PARKINSON'S UK TISSUE BANK AT IMPERIAL COLLEGE LONDON

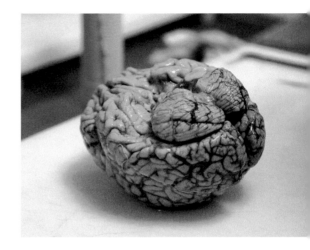

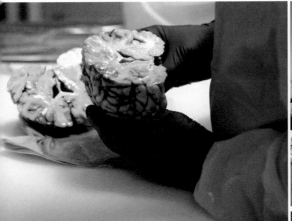

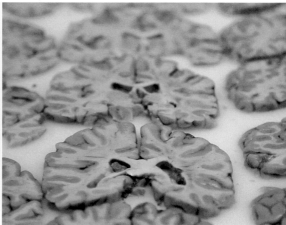

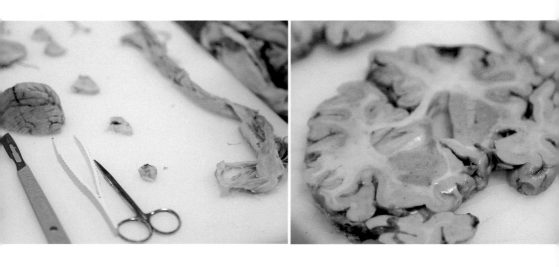

Desiccated half of human brain

c.1800–70
Dry specimen

The brain consists of about 70 per cent water, about 10 per cent more than the human body overall, and therefore shrinks to less than a third of its healthy size when desiccated. Before the widespread use of chemically preserved cadavers or models for medical teaching, anatomical specimens were often dried in the sun, and some respected doctors were known to have used their windowsills for this purpose (though that is unlikely to have been the method of drying for this specimen).

SCIENCE MUSEUM, LONDON

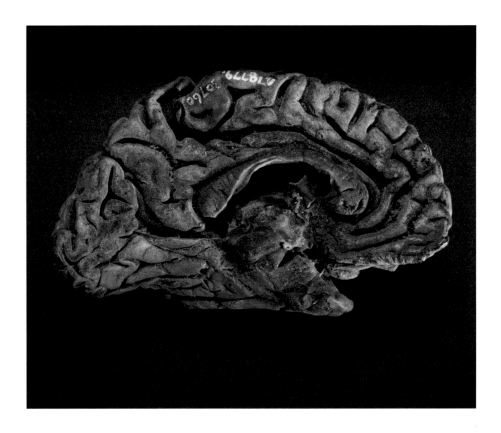

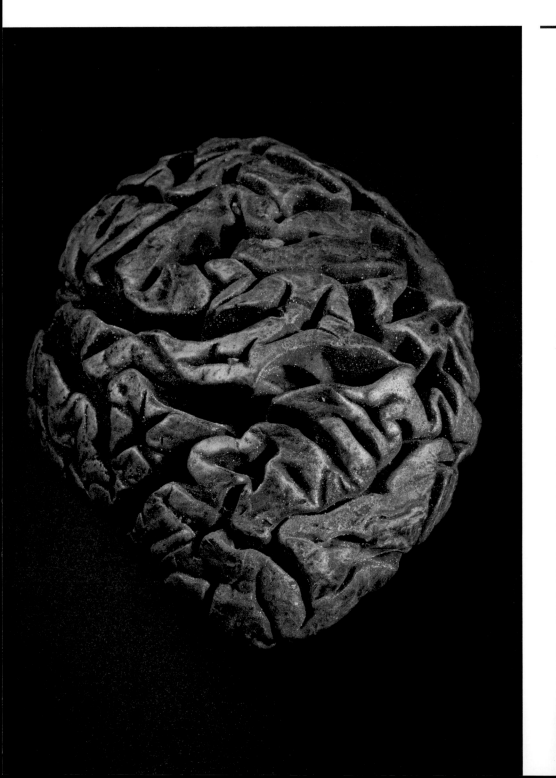

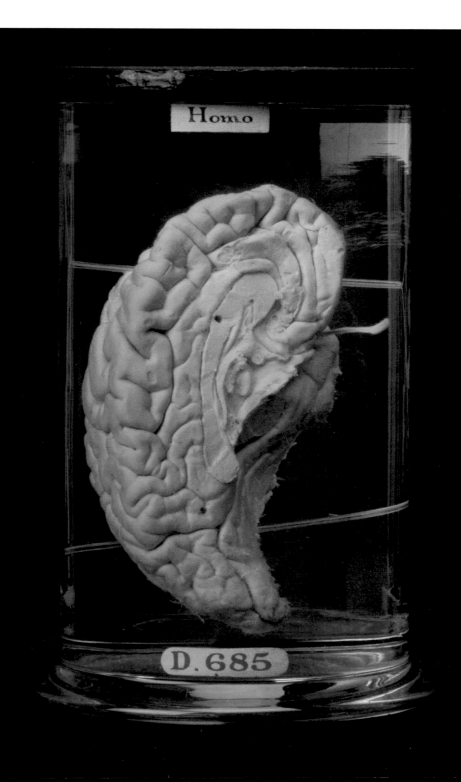

Left hemisphere of the brain of Charles Babbage (left)
1871
Wet specimen

The English mathematician Charles Babbage (1792–1871) is credited with inventing the first mechanical computer, the 'difference engine'. Babbage bequeathed the brain himself and, in a letter accompanying the donation, his son Henry wrote: "I have no objection ... to the idea of preserving the brain ... Please therefore do what you consider best ... [T]he brain should be known as his, and disposed of in any manner which you consider most conducive to the advancement of human knowledge and the good of the human race."

The brain was described by Sir Victor Horsley in the *Philosophical Transactions of the Royal Society of London*, 1908.
HUNTERIAN MUSEUM, ROYAL COLLEGE OF SURGEONS OF ENGLAND, LONDON

Cerebral hemisphere of an ancient Egyptian (below)
*c.*2010 BCE
Dry specimen

This mummified specimen came from the only complete male skull in the burial ground of Dynasty XI in the temple of Mentuhotep II at Deir el-Bahari. It is among the oldest few brain specimens known. The ancient Egyptians famously disregarded the brain and removed it through the nasal cavity when embalming bodies for the afterlife; however, this example suggests the doctrine did not always prevail. At the time of excavation, around 1900, ancient brains were of particular interest to anthropologists and anatomists seeking evidence of human evolution.
HUNTERIAN MUSEUM, ROYAL COLLEGE OF SURGEONS OF ENGLAND, LONDON

Photograph of Einstein's brain
Thomas Harvey, 1955

After Albert Einstein (1876–1955) died, at Princeton Hospital in the USA, his brain was removed at autopsy and kept by pathologist Thomas Harvey. This was against Einstein's express wishes that his remains be cremated and scattered, but his brain had already become an icon of popular culture. Its subsequent journeys, and scientific claims that the source of Einstein's abilities could be detected in either its cellular structure or the pattern of its cortical folds, have become the subject of some controversy.

COURTESY OF DR MARK LYTHGOE, CENTRE FOR ADVANCED BIOMEDICAL IMAGING, UNIVERSITY COLLEGE LONDON

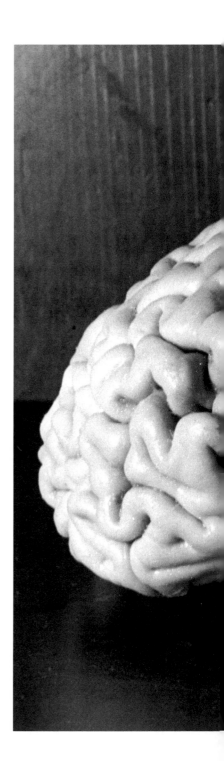

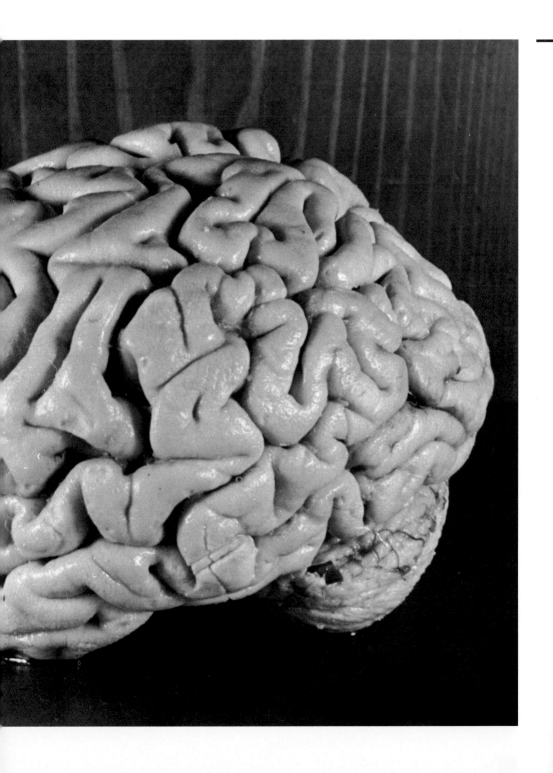

Piece of William Burke's brain
c.1829
Dry specimen

William Burke (1792–1829), with his partner in crime, William Hare, was hanged for the murder of 17 people for the purposes of dissection by the Edinburgh anatomist and lecturer Robert Knox. Their method of smothering victims gave rise to the term 'burking'. In a deliberate piece of poetic justice, Burke's remains were dissected and later exhibited at the University of Edinburgh museum. Souvenirs were made from his skin, as well as, it appears from this neatly packaged specimen, his brain. The case highlighed a crisis in the supply of cadavers for medical education.

SCIENCE MUSEUM, LONDON

Brandenburg State Hospital
Daniel Alexander, 2011
C-type prints

The role of doctors and scientists in the Nazi extermination programmes was recognised by the Nuremberg trials at the end of World War II. However, much evidence was well hidden and it has taken the efforts of later researchers to uncover the extent of the institutionalised mass murder.

Under the 'euthanasia' policy, the Nazis killed at least 120 000 Germans with apparent hereditary illnesses or otherwise deemed to be feeble in mind and body, a drain on society and a danger to racial health. Murder was by neglect and starvation, drug overdoses, and gassing. In over 2000 cases, the brains of the victims were harvested, preserved and examined by respected neuropathologists, including Dr Julius Hallervorden of the Kaiser Wilhelm Institute for Brain Research in Berlin. He had observed the gassing of children at Brandenburg jail, and after the War continued his research using the resulting material.

By the late 1980s the criminal provenance of the remaining brain collections in universities and research institutes could not be denied. Scholars and the families of victims began to trace specimens and give them proper funerals. The brains of young Alfred, Günther and Herbert K (two brothers and a cousin) were returned from Vienna to the mental hospital in Brandenburg's Görden forest, where they had officially died in 1942 and 1944. The specimens had enabled research into a rare condition that affected mental and motor function, named after the scientist who published the paper on them in 1954.

The boys' remains were buried in the hospital cemetery in 2003. During the tidying of the grounds in preparation for this, the wartime graves of twins Werner and Günter Binder, who had also been the subject of a scientific paper in 1942, were uncovered.

This specially commissioned photo essay by Daniel Alexander contemplates some of the key commemorative places at Görden. The site is now a busy modern clinic and we are grateful to patients and staff for their assistance.

Haus 40, exterior (right)

Designed as a private patients' kitchen when the hospital was built around 1910, the house served as a pathology lab by the late 1930s.

Cemetery chapel (right, below)

The Görden hospital has a large secluded cemetery where former patients and staff are buried, as well as other local residents. Many of the graves are now unmarked. 'God Is Love' is inscribed above the door of its chapel, built before World War I.

Haus 40, interior (pages 148–49)

The autopsy rooms were probably located here during the Nazi period, within the pathology department. At the time of photography, the premises were awaiting conversion work and had recently been used for storing furniture.

Children's cemetery: the graves of twins Werner and Günter Binder (1941)

These twins had been diagnosed with hydrocephalus and megalencephaly, causing enlargement of the brain and head, but their official cause of death was pneumonia following scarlet fever. They died on the same day in May 1941, just short of their sixth birthday.

Children's cemetery: the grave of Alfred, Günther and Herbert K (2003)

These boys suffered from a rare hereditary neural disease and were likely murdered in 1942 and 1944, aged seven, two and one.

The preserved parts of their brains were buried at a funeral in 2003 after having been returned from a neuropathology collection in Vienna.

Self-portrait with saw (below)
William Utermohlen, 1997
Oil on canvas

This self-portrait was begun on the day that William Utermohlen (1933–2007) willed his brain to science, after having been diagnosed with Alzheimer's disease in 1995. Alzheimer's, a progressive form of dementia, robs patients of their sense of space, among other symptoms. While there are many observable signs, an Alzheimer's diagnosis cannot be verified for sure until the brain is examined post-mortem. This self-portrait shows the claustrophobic framing of Utermohlen's self-image, with the instruments and targets of his future autopsy, and a symbolic colour scheme presaging death.

GALERIE BECKEL ODILLE BOÏCOS, PARIS

Head I (right)
William Utermohlen, 2000
Pencil on paper

In this, the last of Utermohlen's artworks, the features of the self are eroding and collapsing. Eyes merge with nostrils as two black holes, like a skull, and a deep crack has appeared between the brain and the mouth through which it once communicated.

GALERIE BECKEL ODILLE BOÏCOS, PARIS

Mind Over Matter: The Brain Donors
Ania Dabrowska, 2008–11

'Mind Over Matter' was developed as a joint project between an artist (Dabrowska) and a cultural geographer (Bronwyn Parry). This series of interviews and photographs draws back the veil of secrecy that has surrounded the practice of body donation, revealing the individuals who have elected to donate their brains after death for research, particularly into neurodegenerative conditions such as Alzheimer's disease.

ALL IMAGES COURTESY OF ANIA DABROWSKA AND BRONWYN PARRY

Mrs Lucy Ridsdale (left)

Born in Great Yarmouth in 1910, Lucy Ridsdale remembered things well up to 1936; from then, her memory was gone. She was a headmistress of a boarding school in Uganda and had five children. She always kept her mind active, such as by translating the Gospels into French, Lunyoro and Greek before going to church on Sundays. She died in 2011.

Mr Albert Webb (below left)

Born in Peckham in 1919, Albert Webb worked in print all his life. During the war he served in the army in Naples, where he developed a love for opera. His wife Ellen died of Alzheimer's disease. He is wearing a jumper he knitted himself, depicting his late dog, Lucy. He says of brain donation: "I shall be doing a bit of good perhaps to somebody."

Mr Eddie Holden (below right)

Eddie Holden volunteered for the Parachute Regiment at 17 and took part in the liberation of prisoners of war from the Japanese camps. He would like his brain to be used to discover cures for diseases such as Alzheimer's because he thinks that "losing memory, where you are, not knowing who you are, is a terrible thing".

BRAINS: THE MIND AS MATTER

Mind Over Matter: After I'm Gone
Ania Dabrowska, 2011

Taken at the Brain Bank Laboratory in Cambridge University Hospital, this series of photographs captures the intricate set of processes involved in retrieving, preparing and storing the brain after death. While some of these mortuary scenes may appear stark and impersonal, in fact they represent a profound moment of transaction, one that is shared by all of the donors, regardless of age, race or gender.

ALL IMAGES COURTESY OF ANIA DABROWSKA AND BRONWYN PARRY

Fresh Brain (below left)

A brain freshly removed from a donor's body, before it is 'fixed' by soaking for at least two weeks in preservatives. This image was based on a medical photograph by the Brain Bank Laboratory at Cambridge University Hospital.

Utensils (below right)

Headrest (right)

Showing a headrest on the slab in the dissection room at Cambridge University Hospital, for steadying the head while the brain is removed. The large recess is to hold the neck of an adult, the medium one a child, and the small one a baby.

GIVING/TAKING 157

SCAN
Nina Sellars, 2012

Doctors and scientists are increasingly referring to, and manipulating, detailed scanning imagery as they treat and investigate the brain. The apparent reality of the brain now lies in these intangible, 'dematerialised' forms that exist principally as digital datasets stored in computers.

Australian artist Nina Sellars's 'SCAN' shows a QR (quick response) code that, when read by a smartphone, links to an online animation comprising MRI scans of her brain. These scans were made during treatment for a brain tumour that caused her to lose her sight for a year and a half.

COURTESY OF THE ARTIST AND GV ART, LONDON

Further reading

Abraham C. *Possessing Genius: The bizarre odyssey of Einstein's brain*. Cambridge: Icon Books; 2004.

Bliss M. *Harvey Cushing: A life in surgery*. New York: Oxford University Press; 2005.

Burrell B. *Postcards from the Brain Museum: The improbable search for meaning in the matter of famous minds*. New York: Broadway Books; 2004.

Damasio AR. *Descartes' Error: Emotion, reason and the human brain*. New York: Avon Books; 1994.

DeFelipe J. *Cajal's Butterflies of the Soul: Science and art*. New York: Oxford University Press; 2010.

Gould SJ. *The Mismeasure of Man* (Revised Edition). New York: WW Norton; 1996.

Rose S. *The 21st-Century Brain: Explaining, mending and manipulating the mind*. London: Vintage; 2006.

Sacks O. *The Man Who Mistook his Wife for a Hat*. London: Picador; 1986.

Sawday J. *The Body Emblazoned: Dissection and the human body in Renaissance culture*. London: Routledge; 1995.

Schoonover C. *Portraits of the Mind: Visualizing the brain from antiquity to the 21st century*. New York: Abrams; 2010.

Stafford T, Webb M. *Mind Hacks*. Sebastopol, CA: O'Reilly Media Inc.; 2005.

Zimmer C. *Soul Made Flesh: The discovery of the brain and how it changed the world*. London: Arrow Books; 2005.

Acknowledgements

The authors and editors would like to thank:

At Wellcome Collection:
James Peto, Senior Curator
Jane Holmes, Exhibition Project Manager
Rosalind Leake, Registrar
Rachel Collins, Communications Manager
Marianne Dear, Senior Graphic Designer
Kirty Topiwala, Commissioning Editor
Tom Freeman, Assistant Editor
Tania Moore, Public Programmes Assistant
David Sayer, Photographer
Ben Gilbert, Photographer

Members of the Audio-visual, Conservation, Library, Media, Public Programmes and Publishing teams whose hard work, knowledge and skills made this book and the *Brains* exhibition possible

The Centre for Art, Architecture and Design and the School of Art, Design and Media at the University of Portsmouth.

Special thanks to:

Bettina Beer, Poppy Bowers, Rolf Brüggemann, Brian Burrell, Maria Angeles Ramón y Cajal, Martyn Cooke, Terry Dagradi, Michele-Anne Dauppe, Robert Devčić, David Dexter, Anna Dhody, Barbara Finlay, Edda Gehrmann, Marjan Groot, Simone Gumtau, Friedrich Hauer, Uta Kornmeier, Laura M Linke, Nina Long, Philip Loring, Virginia García Marín, Gene Myers, Adrianne Noë, Sara Perry, Maggie Reilly, Colleen M Schmitz, Sarah Turner, Jenny Walden, Navena Widulin and Rob Zwijnenberg.

And to all those museums, galleries, universities, research institutes and their staff, as well as artists, scholars and private collectors, who have allowed us to show their works.